D1035651

Measuring Physical Activity and Energy Expenditure

Henry J. Montoye, PhD
University of Wisconsin, Madison

Han C.G. Kemper, PhD
Vrije Universiteit, Amsterdam, The Netherlands

Wim H.M. Saris, MD, PhD
University of Limburg, Maastricht, The Netherlands

Richard A. Washburn, PhD
University of Illinois at Urbana-Champaign

Human Kinetics

Library of Congress Cataloging-in-Publication Data

Measuring physical activity and energy expenditure / Henry J. Montoye ... [et al.].
p. cm.
Includes bibliographical references and index.
ISBN 0-87322-500-7 (hbk.)
1. Physical fitness--Measurement. 2. Exercise--Measurement. 3. Energy metabolism--Measurement. I. Montoye, Henry J. (Henry Joseph), 1921- .
QP301.M376 1996
612'.042'0287--dc20 94-38604
CIP

ISBN: 0-87322-500-7

Developmental Editors: Mary E. Fowler and Marni Basic; **Assistant Editors:** Henry Woolsey and Ann Greenseth; **Copyeditor:** Molly Bentsen; **Proofreader:** Anne Meyer Byler; **Indexer:** Theresa J. Schaefer; **Typesetter and Page Layout:** Julie Overholt; **Cover Designer:** Jack Davis; **Illustrator:** Craig Ronto; **Printer:** Braun-Brumfield

Printed in the United States of America 10 9 8 7 6 5 4 3 2 1

Human Kinetics
P.O. Box 5076, Champaign, IL 61825-5076
1-800-747-4457

Canada: Human Kinetics, Box 24040,
Windsor, ON N8Y 4Y9
1-800-465-7301 (in Canada only)

Europe: Human Kinetics, P.O. Box IW14,
Leeds LS16 6TR, United Kingdom
(44) 1132 781708

Australia: Human Kinetics, 2 Ingrid Street,
Clapham 5062, South Australia
(08) 371 3755

New Zealand: Human Kinetics, P.O. Box 105-231
Auckland 1
(09) 523 3462

CONTENTS

Preface vi

Acknowledgments vii

Part I INTRODUCTION AND LABORATORY METHODS 1

Chapter 1 Plan and Scope *3*

Physical Activity 3
Energy Expenditure 4
Assessment Methods 4

Chapter 2 Basic Principles and Laboratory Methods *6*

Measuring Energy Expenditure 6
Biomechanical Methods of Estimating Energy Expenditure 10
Energy Consumption 12

Part II FIELD METHODS OF ASSESSMENT 15

Chapter 3 Doubly Labeled Water *17*

History of the Method 17
Field Procedures 18
Assumptions and Sources of Error 18
Validation of the Method 19
Recent Applications 20
Advantages and Disadvantages 21

Chapter 4 Behavioral Observation and Time/Motion Analyses *26*

Simple Observation Methods 26
Complex Observation Methods 27
Reliability 31

Chapter 5 The Diary Method *34*

Advantages and Disadvantages 34
Validity in Adults 35
Validity in Children 38

Chapter 6 Questionnaires and Interviews *42*

Organizing an Approach 42
Advantages and Limitations 42
Young and Middle-Aged Adults 43
Elderly Adults 54
Children 56

Chapter 7 Movement Assessment Devices *72*

Pedometers 72
Other Step Counters 75
Motion Counters 75
Portable Accelerometers 79

Chapter 8 Estimating Energy Expenditure From Physiologic Response to Activity *97*

Body Temperature 97
Blood Pressure 97
Ventilation 97
Electromyography 98
Heart Rate 98
Multiple Recording Systems to Estimate Energy Expenditure 105
Multiple Methods of Measurement 109

Chapter 9 Conclusions *116*

Final Evaluation Summary 117
Suggestions for Further Research 117

Appendix A Abbreviations, Symbols, and Conversion Table 119

Appendix B Procedures for Measuring Oxygen Consumption ($\dot{V}O_2$): Open Circuit Method 121

Appendix C Classification by Energy Cost of Human Physical Activities 123

Appendix D Tecumseh and Minnesota Occupational and Leisure Time Activity Questionnaires 134

Appendix E Paffenbarger/Harvard Alumni Questionnaire 149

Appendix F Physical Activity Recall Items 152

Appendix G Baecke Questionnaire 154

Appendix H Health Insurance Plan of New York Questionnaire 156
Appendix I British Civil Servant Questionnaire 158
Appendix J Framingham Leisure Time Physical Activity Questionnaire 161
Appendix K U.S. National Health Survey, 1991, Physical Activity Questionnaire 164
Appendix L Modified Baecke Questionnaire 169
Appendix M Zutphen Physical Activity Questionnaire 172
Appendix N The Yale Physical Activity Survey 175
Appendix O Netherlands Health Education Project Questionnaire 181
Appendix P The Amsterdam Growth Study Questionnaire 182
Index 184
About the Authors 191

PREFACE

While most experts agree that remaining active throughout life is important for continued health and fitness, measuring activity levels outside of the laboratory is difficult. *Measuring Physical Activity and Energy Expenditure* brings together the diverse literature on physical activity assessment, providing investigators, clinicians, and educators with state of the art, scientific methods for assessing physical activity and energy expenditure. Using better physical activity assessment techniques in research will clarify how physical activity relates to the prevention and treatment of diseases and other health-related conditions.

You'll find *Measuring Physical Activity and Energy Expenditure* to be a comprehensive source on physical activity and energy expenditure measurement in both field and laboratory settings. We have attempted to review and evaluate the many approaches currently available for measuring physical activity and energy expenditure and present empirical evidence supporting the use of these techniques. We've also endeavored to highlight the research necessary for future improvements in the validity and precision of methods of estimating habitual physical activity and energy expenditure.

This book is divided into two parts. In Part I we introduce the concepts of physical activity and energy expenditure and describe the physiological and biomechanical basis for measuring energy expenditure in the laboratory. In Part II we discuss physiological, behavioral, and biomechanical techniques for measuring energy expenditure or physical activity in field research. We review the vast amount of research on the measurement of energy expenditure and physical activity, emphasizing the validity, reproducibility, and feasibility of these procedures. We then offer conclusions about where the science of estimating energy expenditure and physical activity in humans now stands. Pragmatic information is provided including suggestions about the use of equipment, in particular for field assessment of physical activity and energy expenditure. The appendixes feature a number of questionnaire and interview forms that represent the most widely used resources available for this approach.

The contributors to *Measuring Physical Activity and Energy Expenditure* have had extensive experience in the topic and are developing innovative approaches to field assessment. The objective has been to make this book practical and relevant, not theoretical. We hope you'll use the information in this book to help make significant contributions to the health of people everywhere.

Henry J. Montoye
Han C.G. Kemper
Wim H.M. Saris
Richard A. Washburn

ACKNOWLEDGMENTS

We are indebted to many people who were helpful in developing this handbook. Appreciation is expressed for permission to reproduce the questionnaires and other materials. The assistance of the publisher in bringing the handbook to its final form is acknowledged. We are especially grateful to Ms. Phyllis Sierra for her help and her superb typing of the first draft and revisions of the manuscript. She seemed to have developed as much interest in the project as the authors.

Part I

Introduction and Laboratory Methods

Chapter 1

PLAN AND SCOPE

In today's world, chronic diseases and disabilities account for most of the ill health among residents in highly developed countries. Most of these contemporary health problems (including coronary heart disease, obesity, stress, hypertension, non-insulin-dependent diabetes, osteoporosis, some types of cancer, and emotional disorders) appear to be associated with our habits of living, including low levels of physical activity. This association has been mainly responsible for the surge in research during the last 40 years to define the relationship of physical activity to health status. To some extent it has also resulted in an increase in regular exercise of at least some individuals in recent years. To refine our understanding of the association between physical activity and health, we need better methods for assessing habitual physical activity and energy expenditure in the daily lives of men, women, and children. This need was emphasized in the Report of the Working Group on Heart Disease Epidemiology (1979) of the National Institutes of Health; at a 1984 conference at the Centers for Disease Control, Atlanta, Georgia, USA (Caspersen, Powell, & Christenson, 1985; LaPorte, Montoye, & Caspersen, 1985; Stephens, Jacobs, & White, 1985); and in an extensive report of the National Center for Health Statistics (Drury, 1989). The desire for better methods for assessing children's physical activity was expressed at a meeting of pediatric work physiology (Bell, Maceh, Rutenfranz, & Saris, 1986).

Physical Activity

An important aspect of any assessment of habitual physical activity is the definition and interpretation of the term *physical activity*. Because human beings obey the law of the conservation of energy and must fuel all activity by extracting energy from food, measurements of physical activity are often expressed in terms of energy expenditure. Alternately, physical activity can be expressed as the amount of work performed (watts), as the time period of activity (hours, minutes), as units of movements (counts), or even as a numerical score derived from responses to a questionnaire. Activity can also be defined as an overt intentional behavior. The number of social contacts, for example, is one result of studying physical activity in this way. It is clear, then, that any particular assessment technique measures only one part of so-called ''habitual physical activity.''

Physical activity is commonly described as having three dimensions: duration (e.g., minutes, hours), frequency (e.g., times per week or per month), and intensity, or strenuousness (e.g., rate of energy expenditure in kilocalories per minute or kilojoules per hour). However, there is a fourth dimension often overlooked, namely, the circumstances and purpose of the activity. Both physical environment (ambient temperature, altitude, etc.) and psychological or emotional circumstances may modify the physiological effects of an activity. A muscle does not know whether it is swinging a hammer or a tennis racket. But, voluntarily shoveling snow on a pleasant winter evening may well have a different physiological effect on the body than hurriedly shoveling out the driveway in the morning to get to work on time. Joseph B. Wolffe, a cardiologist and first president of the American College of Sports Medicine, expressed this thought as follows: When a dog chases a fox through a field, both may be traveling at the same rate of speed, but the physiological effects on the fox may be expected to be different than the effects on the dog (personal communication, 1962).

It is tempting to assume that technology has eliminated significant physical activity from most occupations in our industrialized societies. Over the years, mechanization has decreased the human energy required in most occupations, including agricultural work. For the many people with completely sedentary occupations, leisure time activity, including going to and from the job, is the only source of exercise. But when we view a total community of workers, we find that physical activity required on the job is very significant for classifying people on the basis of total 24-hr physical activity. This has been clearly shown among adult males in a total community in the United States by Reiff and others (1967) and in India by Mayer, Roy, and Mitra (1956). In general, one can say that the duration of an activity contributes more to the total daily activity score than its intensity.

Energy Expenditure

The term *energy expenditure* is not synonymous with physical activity or exercise. One may expend the same amount of energy in a short burst of strenuous exercise as in less intense endurance-type activity, but the health and physiological effects of the two could be different. For example, it is generally accepted for adults that high-intensity exercise carried on for a short period improves maximal oxygen uptake ($\dot{V}O_2max$) more than moderate exercise over a longer period.

One observation about energy expenditure is essential to keep in mind. The intake or expenditure of joules is related to body size. A small person who is very active may expend a similar number of kilojoules in 24 hr as a large person who is sedentary. So if exercise is to be expressed as energy expenditure in joules or calories, body size must be taken into account. To this end, energy expended or ingested is sometimes given as kilojoules or kilocalories per unit of body weight or, in the case of oxygen uptake, as milliliters of O_2 per kilogram of body weight. The use of METs (an abbreviation of metabolic) is another approach to correcting for body weight. A MET represents the ratio of energy expended in kilojoules (kilocalories) divided by resting energy expenditure in kilojoules (kilocalories), either measured or estimated from body size. In estimating resting (not basal) energy expenditure, a value of 4.2 kJ (1 kcal) per kilogram of body weight per hour or 3.5 ml O_2 utilized per kilogram of body weight per minute gives reasonably satisfactory results in most cases. Although neither method is perfect, the MET approach is more popular and probably more useful. Although he did not use the term MET, LaGrange (1905) almost a century ago expressed the strenuousness of activities as a ratio of exercise metabolism to resting metabolism. The World Health Organization adopted the same principle in its physical activity index. Among exercise physiologists, it is almost universally accepted to use METs to express energy expenditure in relation to body weight.

Energy is expended in three ways in warm-blooded humans and animals. A certain amount of energy is required at rest to maintain body temperature and involuntary muscular contraction for functions including circulation and respiration. This energy level represents the resting metabolic rate. Second, some energy is required to digest and assimilate food. This process, formerly called *specific dynamic action* and now referred to as *dietary induced thermogenesis* or *thermic effect of food,* adds about 10% to the resting metabolic rate. These two represent but a small part of the total energy expenditure and can be altered only very slightly in individuals. By far the most important source of variation between individuals in energy expenditure (when adjusted for body size) is the muscular activity carried out. The sources of this activity are one's daily work, leisure pursuits, and transportation to and from work or other destinations (which some investigators include as part of leisure time activity).

In the International System of Units (SI), the unit of measurement for heat production is the joule (named for James Prescott Joule,who did pioneering work in metabolism). However, the calorie unit (the energy required to heat a gram of water 1 °C) is also used. In this book, we will use joules with equivalent calories generally shown in parentheses. The large calorie, or kilocalorie (kcal), is equal to 1,000 calories, just as the kilojoule (kJ) is equal to 1,000 joules. When human diet is discussed (in connection with weight loss programs, for example), the large calorie is ordinarily used. Sometimes the British thermal unit (BTU) is employed. These units can be easily converted as follows (see Appendix A for additional conversions):

$$1 \text{ kJ} = 0.238 \text{ kcal} = 0.948 \text{ BTU}$$

Assessment Methods

The greatest obstacle to validating field methods of assessing habitual physical activity or energy expenditure in humans has been the lack of an adequate criterion to which techniques may be compared. The practice of intercorrelating various field methods may be of some value, but because there are errors in all methods it is impossible to determine the true validity of any one of them in doing so. Furthermore, a difference in units of the outcome measure does not readily permit validation based on an intercorrelation of methods. For example, $kJ \cdot hr^{-1}$ is not comparable to counts measured on a movement counter. As an alternative criterion measure, investigators have employed work capacity or aerobic power ($\dot{V}O_2max$) or a change in these parameters. More active subjects tend to have greater aerobic power. However, measures of physical fitness and the ability to improve in fitness with training have a sizable hereditary component (Bouchard, 1986; Bouchard, Boulay, Simoneau, Lortie, & Pérusse, 1988) and are affected by gender, aging, relative weight, and habits other than physical activity (Leon, Jacobs, DeBacker, & Taylor, 1981).

Fitness measures thus leave much to be desired as criteria for validating methods of assessing habitual physical activity. Furthermore, the kind of exercise that increases aerobic power may not be the only or even the most important physical activity related to health and disease. A relatively new technique, doubly labeled water, which can provide an integrated measure of energy expenditure over time, is a potentially acceptable gold standard for the validation of other methods of assessing energy expenditure. If the doubly labeled water technique (described in detail in chapter 3) proves to be an effective criterion, it will be useful only in validating measures of overall energy expenditure.

SUMMARY

Validated methods of estimating habitual physical activity are needed to study further the relationship of regular exercise to health. Physical activity can be described in terms of duration, frequency, strenuousness, and the circumstances surrounding the activity. If total activity or energy expenditure is to be assessed, both occupational and leisure time activities must be considered. Because a large person can be expected to expend more energy than a small person, it is necessary to adjust for body size if energy expenditure (kJ or kcal) is to be interpreted as physical activity. The concept of METs is the most useful way of doing this.

REFERENCES

Bell, R.D., Maceh, M., Rutenfranz, J., & Saris, W.H.M. (1986). Health indicators and risk factors of cardiovascular diseases during childhood and adolescence. In J. Rutenfranz, R. Mocellin, & F. Klint (Eds.), *Children and exercise XIII* (pp. 59-88). Champaign, IL: Human Kinetics.

Bouchard,C. (1986). Genetics of aerobic power and capacity. In R.M. Malina & C. Bouchard (Eds.), *Sport and human genetics* (pp. 59-88). Champaign, IL: Human Kinetics.

Bouchard, C., Boulay, M.R., Simoneau, J.A., Lortie, G., & Pérusse, L. (1988). Heredity and trainability of aerobic and anaerobic performance: An update. *Sports Medicine*, **5**, 69-73.

Caspersen, C.J., Powell, K.E., & Christenson, G.M. (1985). Physical activity, exercise, and physical fitness: Definitions and distinctions for health-related research. *Public Health Reports,* **100**, 126-130.

Drury, T.F. (Ed.) (1989). *Assessing physical fitness and physical activity in population-based surveys* (DHHS Pub. No. [PHS] 89-1253). Washington, DC: U.S. Government Printing Office.

LaGrange, F. (1905). *Physiology of bodily exercise.* New York: D. Appleton.

LaPorte, R.E., Montoye, H.J., & Caspersen, C.J. (1985). Assessment of physical activity in epidemiologic research: Problems and prospects. *Public Health Reports,* **100**, 131-146.

Leon, A.S., Jacobs, D.R., Jr., DeBacker, G., & Taylor, H.L. (1981). Relationship of physical characteristics and life habits to treadmill exercise capacity. *American Journal of Epidemiology,* **113**, 653-660.

Mayer, J., Roy, P., & Mitra, K.P. (1956). Relation between caloric intake, body weight and physical work in an industrial work population in West Bengal. *American Journal of Clinical Nutrition,* **4**, 169-176.

Reiff, G.G., Montoye, H.J., Remington, R.D., Napier, J.A., Metzner, H.L., & Epstein, F.H. (1967). Assessment of physical activity by questionnaire and interview. *Journal of Sports Medicine and Physical Fitness,* **7**, 135-142.

Report of the Working Group on Heart Disease Epidemiology (1979). (NIH Publ. No. 79-1667). Bethesda, MD: National Heart, Lung, and Blood Institute.

Stephens, T., Jacobs, D.R., Jr., & White, C.C. (1985). A descriptive epidemiology of leisure-time physical activity. *Public Health Reports*, **100**, 147-157.

Chapter 2

Basic Principles and Laboratory Methods

The physiological and biomechanical principles underlying movement and energy expenditure are complex, and numerous difficulties can be encountered in developing simple field techniques for assessing habitual activity. For the most part, laboratory methods are not useful in the field for measuring activity and energy expenditure. The methods we describe in this chapter do not directly apply to epidemiologic studies of assessing habitual physical activity. But it is important to understand the basis of these laboratory techniques for several reasons:

- They are generally used to determine the energy cost of specific activities.
- They are useful in some studies of limited scope.
- They frequently have served as criteria for validating simpler, practical field methods.

In the middle of the 18th century, Lavoisier conceived the first law of thermodynamics, that energy can be neither created nor destroyed but only changed from one form to another. This principle of the conservation of energy was later formulated by Mayer in 1842 and Helmholtz in 1847, but it remained for Joule, a brewer, to provide experimental data to support the concept (Fenn & Rahn, 1964). When Lavoisier and Laplace demonstrated that muscular exercise consumes oxygen and produces carbon dioxide (Chapman & Mitchell, 1965), the stage was set for learning how to measure energy expenditure. It was clear then that the energy in the food consumed should equal the energy expended, provided of course the stored energy in the individual remained constant.

Measuring Energy Expenditure

The direct measurement of energy expenditure (heat production) by a living animal or human being is possible. Although the engineering problems are formidable, the heat produced while the subject is in a sealed, insulated chamber can be measured.

Heat Production—Calorimetry

Probably the earliest calorimeter was the one devised by Joseph Black in 1761, which consisted of a block of ice into which was placed a substance of known mass and temperature (Perkins, 1964). The heat produced was determined from the amount of water produced as the substance gradually cooled to the temperature of the ice. From 1782 to 1784 Lavoisier collaborated with the mathematician Laplace to measure the heat production of living animals using a modification of Black's device (Perkins). Later they measured oxygen consumption and carbon dioxide production in humans. In fact, Lavoisier coined the term *oxygen* for the gas discovered earlier by Joseph Priestly. What other discoveries might we have expected from Lavoisier had he not become a victim of the guillotine during the French revolution?

The modern era in the use of room calorimeters began with the remarkable experiments of Atwater and Benedict from 1896 to 1902 (Atwater & Benedict, 1903; Garrow, 1978). In the past few decades, several room calorimeters have been used for research on energy metabolism. Webb (1985) gives a detailed description and history of this technique.

A room calorimeter measures the heat produced by the subject at rest or during exercise by circulating water through pipes in the insulated chamber and carefully measuring, at frequent intervals, the temperature of the ingoing and outgoing water and the water flow. Sophisticated engineering is required to prevent heat loss from the chamber by other means. The latent heat of the water vaporized must be determined by measuring the vapor in the ventilating air current. Calorimeters have been built in which air flow and temperature are measured by means of thermocouples using the thermal gradient principle (Carlson & Hsieh, 1970; Jéquier, Acheson, & Schutz, 1987). Energy exchange during muscular exercise can be measured by installing an exercise device (treadmill, bicycle ergometer, etc.) in the chamber.

Webb (Webb, 1980; Webb, Annis, & Troutman, 1980) also describes an insulated, water-cooled suit worn by the subject in which the flow of water through the suit and the temperature of the incoming and outgoing water are measured to determine heat production. The suit has been modified slightly by Hambraeus, Forslund, & Sjödin (1991). A subject can also exercise while wearing the insulated suit (see Figure 2.1). In fact, by using the suit as a respirometer or indirect calorimeter, one can measure external work done during cycling on an ergometer or walking on a treadmill (Webb, Saris, Schoffelen, van Ingen Schenau, & ten Hoor, 1988).

Oxygen Consumption—Respiration Chamber

When energy is transformed from food to heat and muscular work, oxygen is consumed and thus the oxygen consumed could be measured to ascertain energy expenditure. Again, the engineering problems are formidable but not insurmountable. The terms *oxygen uptake*, *oxygen intake*, and *oxygen utilization* are generally used synonymously with *oxygen consumption,* and that is the practice in this book. Oxygen consumption is defined as the oxygen taken into the body and used in the tissues. The symbol VO_2 is usually used to express the volume of oxygen; the rate of oxygen utilization per minute is shown as $\dot{V}O_2$. The term *indirect calorimetry* is applied to the method of estimating energy expenditure from VO_2 consumption and VCO_2 production because heat production is not measured directly.

A room calorimeter can be constructed in which expired air is analyzed to estimate heat production. Atwater and Benedict (1905) showed that by measuring the oxygen consumed and carbon dioxide produced, heat production could indeed be estimated with reasonable accuracy. This kind of calorimeter is usually referred to as a respiration chamber.

If one wishes to express the energy expended in kilojoules (kilocalories), it must be recognized that the kilojoules (kilocalories) of heat produced by the utilization of 1 liter (L) of oxygen varies with the foodstuffs consumed. The combustion of 1 L of oxygen yields 19.59 kJ (4.68 kcal) from fat alone, 18.75 kJ (4.48 kcal) from protein alone, and 21.18 kJ (5.06 kcal) from carbohydrate starch alone. Even this is not precise because within each of these three main food sources, the kilojoules (kilocalories) of heat from 1 L of oxygen can vary. For example, considering different types of macronutrients, Brody (1945) gives 18.4 kJ (4.4 kcal) for cottonseed oil and corn oil, 19.3 kJ (4.6 kcal) for butterfat, 21.18 kJ (5.06 kcal) for starch, and 21.26 kJ (5.08 kcal) for sucrose. Similarly, the production of heat from 1 L of carbon dioxide varies with the foodstuffs metabolized. For precise conversion of oxygen utilization to energy expenditure, the proportions of fat, carbohydrates, and protein being metabolized must be known. The proportion of protein being utilized can be determined by the nitrogen that appears in the urine during the time of observation. About 1 gram (g) of nitrogen is excreted for every 6.25 g of protein metabolized.

The ratio of the volume of carbon dioxide produced to the volume of oxygen consumed, the so-called *respiratory quotient (RQ),* gives a reasonable approximation of the percentage of carbohydrate and fat being burned, the ratio being 0.7 when pure fat is the source of energy and 1.00 when it is pure carbohydrate. These ratios assume a ''steady state,'' which exists when the oxygen uptake equals the oxygen requirement of the tissues and there is no accumulation of lactic acid. Heart rate, ventilation, and cardiac output remain at fairly constant levels during a steady state. RQ is not representative of the foodstuffs being oxidized in a non-steady state, such as at the start of exercise or during the onset of acidosis or alkalosis as may occur during strenuous exercise or some disease states. The term *respiratory exchange ratio (RER)* rather than RQ is used when a steady state does not exist.

Variations in the caloric equivalents of different fat, different carbohydrate, and different protein sources can be ignored because the error produced is very small. This is because in a normal diet the mixture of different types of fat, carbohydrate, and protein balances out the differences in caloric equivalents. Even the error

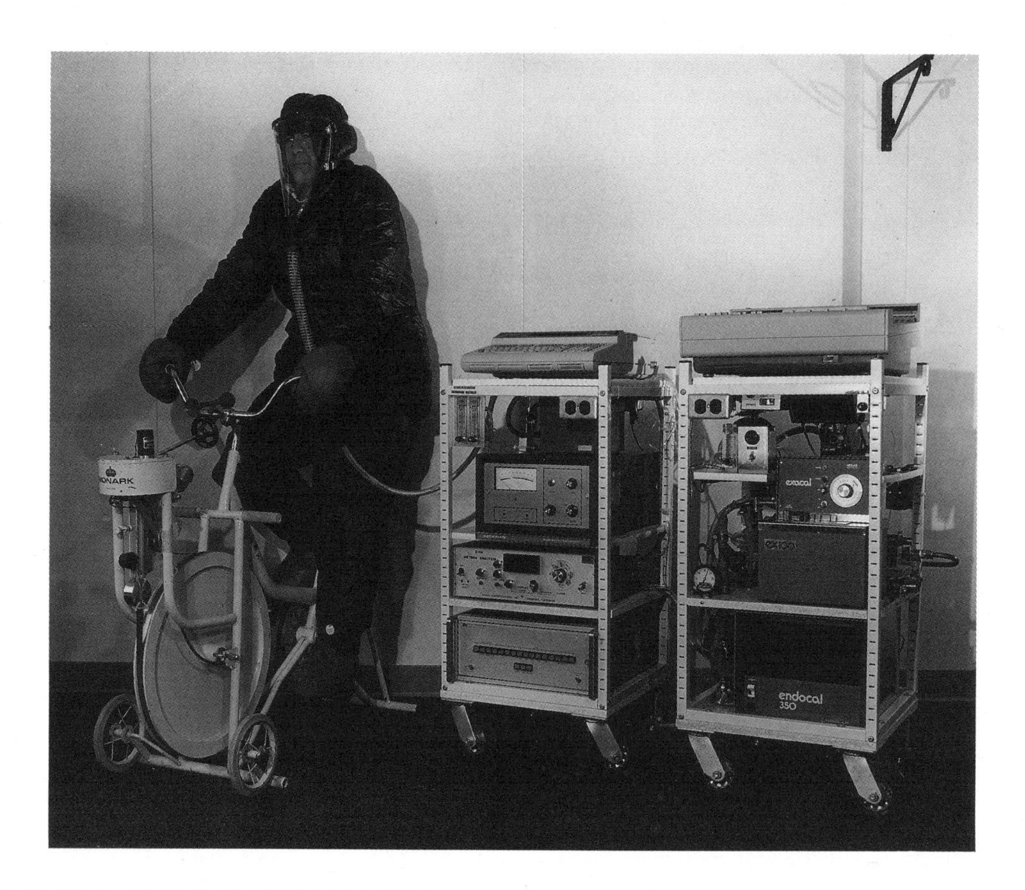

Figure 2.1 Direct calorimeter and metabolism apparatus being used to monitor subject riding a stationary bicycle.

introduced by not measuring the percentage of protein being used can be ignored in most instances because the caloric equivalents of oxygen are similar for carbohydrates and protein. No matter how diverse the actual composition of the food oxidized, the error in estimating energy expenditure is unlikely to be more than 2% to 4%. An error of 100% in the estimation of urinary excretion of nitrogen leads to only a 1% error in energy expenditure. When protein utilization is not considered, the equation then becomes (Weir, 1949):

$$\text{heat production in kcal} = VO_2\ (3.9 + 1.1\ RQ).$$

Oxygen Consumption—Other Methods

Calorimeter rooms and respiration chambers are expensive to build and operate. They are also confining and not suitable for measuring energy expenditure in some activities. But they have been and still are used to validate other methods of estimating energy expenditure.

There are several simpler techniques for measuring oxygen uptake. One, called the *closed circuit method,* requires the subject to be isolated from outside air. The respirometer originally contains pure oxygen, and as the subject breathes in this closed system the carbon dioxide is continuously removed as it passes through soda lime. The gas volume gradually decreases, and the rate of decrease is a measure of the rate of oxygen consumption. Regnault and Reiset developed this system in 1849, and by measuring the carbon dioxide absorbed they discovered the respiratory quotient (Fenn & Rahn, 1964). This method works reasonably well for measuring resting or basal metabolic rate, but absorbing the large volume of carbon dioxide produced during prolonged, strenuous exercise becomes a problem. The *open circuit method* described next is more suited to measuring exercise metabolism.

Two procedures in the open circuit method have been developed. In one, the flow-through technique (Kinney, 1980), a large volume of the equivalent of outside air passes through a hood worn by the subject. The subject inspires and expires into the airstream flowing through the hood. Air flow and percentage of oxygen and carbon dioxide are precisely measured to calculate $\dot{V}O_2$ and RQ. It is necessary to have accurate gas analyzers, particularly the one for carbon dioxide because its concentration may be between 0% and 0.5%. This method is especially useful for long-term measurements with the subject at rest or doing only mild exercise.

The second procedure, the time-honored Douglas bag method (although a Douglas bag may not necessarily be used), has been found to be accurate and theoretically sound. With this procedure, the subject generally wears a nose clip and mouthpiece or a face mask. Outside air or its equivalent is inhaled through the mouthpiece or mask containing a one-way valve and exhaled into a Douglas bag or Tissot tank. It is important that the mouthpiece and connected tubing provide minimal resistance to airflow, or the cost of breathing will increase the energy expenditure. The volume of air in the bag or tank is measured to calculate ventilation. A sample of exhaled air is obtained to measure the O_2 and CO_2 concentrations. This is usually done with a Haldane, modified Haldane, or Micro-Scholander apparatus (see Figure 2.2). These techniques use reagents to absorb the carbon dioxide and oxygen, respectively, with the volume of the sample measured before and after the gases are absorbed.

It is not necessary to measure the volume of both the inspired and expired air because, by subtraction, the percentage of nitrogen (including a small percentage of inert gas) in the exhaled air can be measured. With this information and knowing 1-min ventilation, one can calculate the volume of exhaled nitrogen and, assuming no retention of nitrogen in the body, the volume of inspired air. Because gas volumes are usually expressed in standard conditions (standard temperature, pressure,

Figure 2.2 The measurement of O_2 and CO_2 gas concentrations in a sample of exhaled air by the Micro-Scholander apparatus.

dry; or STPD), barometric pressure, humidity, and temperature of the gases must be determined. When RQ is known, oxygen consumption can be converted to energy expended or an assumption made that on a fixed diet, 1 L of oxygen is roughly equivalent to 20.9 kJ (5 kcal). Details of the methods are given in Appendix B as well as in many exercise physiology texts.

In the laboratory, modern electronic equipment usually replaces the Douglas bag and chemical analyzers, whereby ventilation and oxygen and carbon dioxide percentages are determined instantaneously and continuously (see Figure 2.3). Chemical analyzers are generally used to analyze standard gas mixes to calibrate the electronic equipment. The electronic equipment confines the procedure to laboratory or clinic. The Douglas bag method is not as restricting because a bag can be carried on the back or by an assistant close by. This method thus can be used in the field, but because its use is limited to a few subjects, tested under controlled conditions, we describe it here as a laboratory method.

Nathan Zuntz (1847-1920) recognized the advantage of having the subject carry a self-contained unit (Figure 2.4) if $\dot{V}O_2$ is to be measured during exercise. He developed what was probably the first such unit, which resembled a large rucksack (Zuntz & Leowy, 1909). This was a forerunner of the portable calorimeter designed by Kofranyi and Michaelis (1940). Improvements were made during the subsequent 10 years, resulting in the model by Müller and Franz (1952). This also resembles a rucksack but is smaller and lighter than Zuntz's apparatus.

The Müller-Franz calorimeter registers ventilation and siphons off a small percentage of the expired air into a small attached bag for later analysis. This apparatus functions reasonably well during rest or moderate exercise. At airflows of about 80 to 100 L per minute, the meter begins to underrecord ventilation (Insull, 1954; Montoye, VanHuss, Reineke, & Cockrell, 1958; Orsini & Passmore, 1951) and hence underestimate energy expenditure. At severe exercise, where instantaneous flows can reach 200 L or more per minute, the instrument seriously underestimates energy expenditure. There is also a potential error due to diffusion of the gas through the bag, which becomes more serious the longer the delay in analyzing the gas. In addition to these limitations, there may be some interference in particular activities (the calorimeter weighs about 3 kg), although the instrument can be carried in a bicycle basket or by an assistant. Also, the rates of energy expenditure are averaged over the entire collection period.

Wolff (1958) improved the Kofranyi-Michaelis respirometer. His integrating motor pneumotachograph (IMP) is available from J. Langham Thompson Ltd., Bushey Heath, Hertz, England. The IMP has some of the limitations of the Kofranyi-Michaelis respirometer.

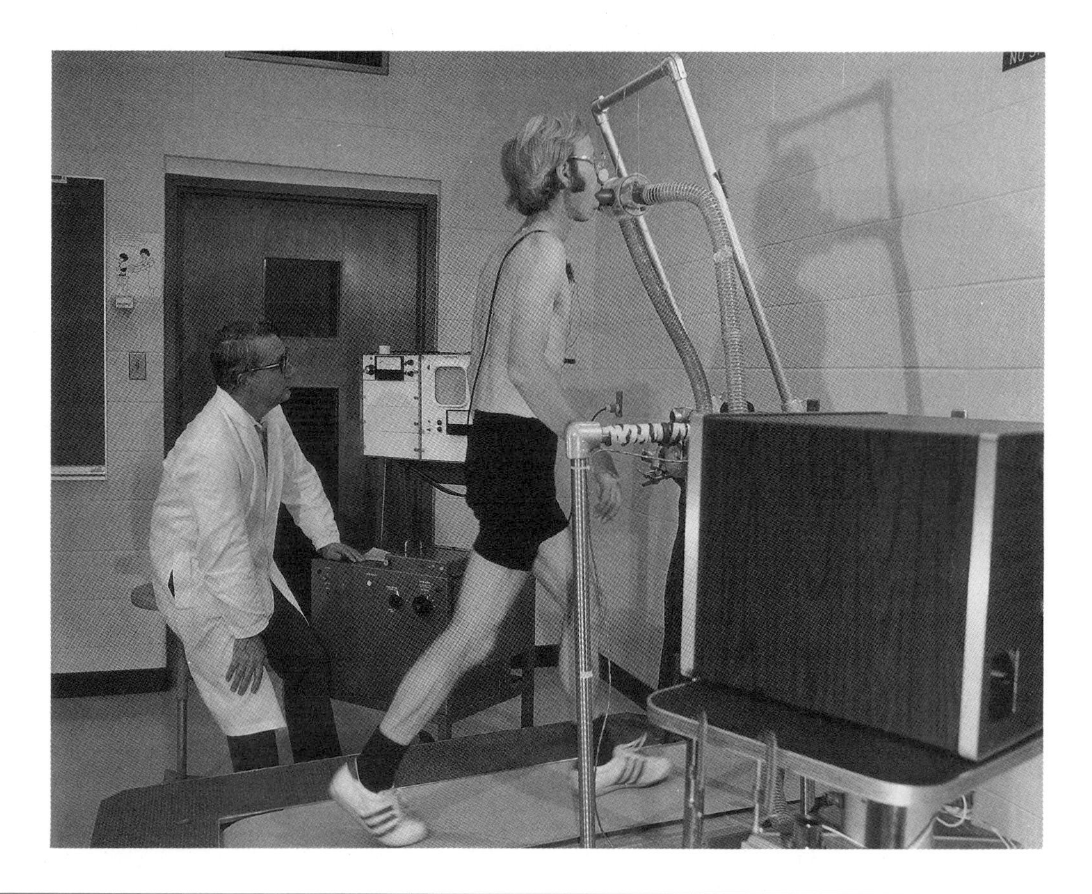

Figure 2.3 The measurement, using electronic instruments, of oxygen consumption during exercise in the laboratory.

Ventilation is integrated electrically rather than mechanically lowering the expiratory resistance. Also, smaller percent samples are possible. This group (Humphrey & Wolff, 1977) later developed a more advanced instrument, the oxylog, available from P.K. Morgan Ltd., Rainham, Kent, England. This battery-operated, self-contained, portable instrument weighs about the same as the Kofranyi-Michaelis respirometer, but it is engineered for on-line measurement of oxygen consumption. Carbon dioxide is not measured so the Weir (1949) formula is employed, using an RQ estimated at 1.0. It has been found to be reasonably accurate in field measurements during rest and up to moderately strenuous exercise (Collins, Abdel-Rahman, & Awad El Karim, 1988; Harrison, Brown, & Belyasin, 1982; McNeill, Cox, & Rivers, 1987). The error was reported to be 2% to 3% at 4 METs, but the error increases at lower and higher workloads (Patterson & Fisher, 1979). Ikegami, Hiruta, Ikegami, and Miyamura (1988) added a telemetry capacity to the oxylog so $\dot{V}O_2$ could be recorded remotely at 1-min intervals.

Biomechanical Methods of Estimating Energy Expenditure

All of the techniques discussed thus far have to do with measuring energy expenditure. As we mentioned previously, muscular activity (exercise) is responsible for most of the variation in energy expenditure among people after correction for body weight, and often it is the activity itself that is of most interest. Several methods are available to record muscular activity and to estimate energy expenditure from data thus obtained.

Photography

One approach is cinematography—to photograph an exercising individual with a high-speed movie camera. This enables measurement of the velocity and acceleration of trunk and limbs. Then, by measuring or estimating the masses of the body segments, one can estimate the energy required to execute the movement. Sophistication in cinematography has increased greatly in recent years. Modern cameras have ranges from a few to more than 10,000 frames per second. It is possible to obtain three-dimensional representations of the activity by simultaneously photographing the subject with two or more cameras.

Before computers were available, film analysis was done by projecting the film, hand-tracing the projections, and then calculating point displacements, velocities, accelerations, and the like—a time-consuming and expensive procedure. Now, it is possible to "digitize" (i.e.,

Figure 2.4 A portable respirometer to measure energy expenditure under field conditions.

plot *x* and *y* coordinates) electronically. This digitizing equipment can be coupled to a microcomputer and printer, with programs available that use the coordinates generated to produce the desired calculations.

When videotaping equipment first came into use about 20 years ago, the resolution (sharpness) was poor and the exposure time was not brief enough for careful quantitative work. However, the technology of videotaping and computer-assisted analysis has improved to the point that videography rivals cinematography for scientific work. The reader interested in more information about these procedures should consult a good text in biomechanics (like the one by Kreighbaum and Barthels, 1990).

Although displacement, acceleration, and velocity of the body and body segments can be measured by photographic and video equipment, the interpretation of these measurements in terms of energy expenditure is complex. Williams (1985) discusses this at length.

Force Platform

Another approach (albeit more confining than photography) in analyzing movement by applying the laws of physics is through the use of a *force plate* or *force platform*. The force plate consists of a platform mounted on a solid foundation and containing sensing elements placed strategically beneath the surface so that forces in three planes may be recorded. Although there may be only a few microns of compression or movement, weight from a fraction of an ounce to a ton or more may be recorded. With the subject on the platform, the system is balanced to zero. Then, as the subject executes a movement or resists an external force as a result of muscular contraction, the sensing elements are subjected to variations in pressure. These are proportional to the forces applied by the body to the platform, equal and opposite reactions to the effort necessary to execute the movement. These dynamic vertical, frontal, and transverse forces are amplified and recorded in the form of continuous curves with a time base. From these data, energy expenditure may be estimated.

The history of three-dimensional force plate measurements goes back at least to the mechanical device of Elftman (1938), who used linear springs as sensors and levers to show forces. Later, Cunningham and Brown (1952) used strain gauges to measure forces. A short time after that, force plates were described employing piezoelectric crystals (Lauru, 1954, 1957) or linear variable differential transformers (Barany, Ismail, & Manning, 1965; Greene & Morris, 1958) as the sensing elements.

Force plates have been used mostly in biomechanical studies; few reports have appeared relating force plate measurements to $\dot{V}O_2$. Lauru (1957) was probably the first to compare $\dot{V}O_2$ with force plate measurements. Brouha (1960) reported a correlation coefficient of .83 between $\dot{V}O_2$ and the integrated vertical force-time curves while three subjects raised and lowered their arms at various rates, with and without hand weights. Although the correlation coefficient under these circumstances is difficult to interpret, the plot (Brouha) resembles the results that Montoye, Servais, and Webster (1986) reported (see Figure 2.5). In the second study, 21 subjects performed three different exercises on the force plate. The standard error for estimating oxygen consumption from the triaxial vector sum was 6.9 ml O_2 per kilogram of body weight per minute.

Ismail and colleagues (Ismail, 1968; Ismail, Barany, & Smith, 1970) mounted a force plate under the bed of a motor-driven treadmill. They measured triaxial forces generated and $\dot{V}O_2$ while subjects walked at speeds from 0.5 to 3 mph. $\dot{V}O_2$ was measured and multiple regression equations were developed using force measurements and body size to predict $\dot{V}O_2$. It is

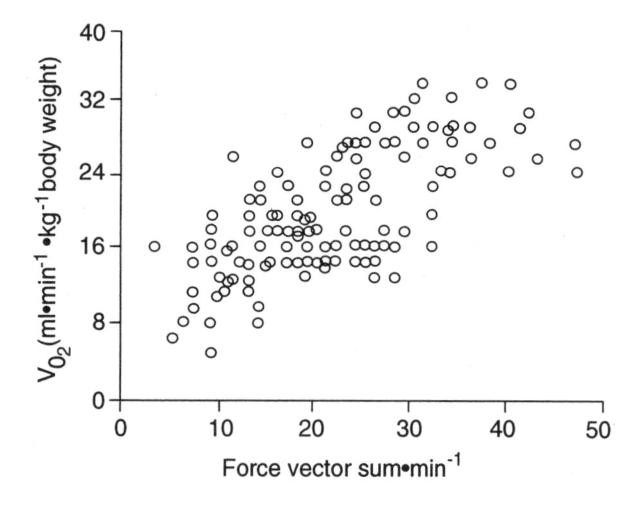

Figure 2.5 The relationship of O_2 and force vector sum · min^{-1}, $S_{y.x}$ = 6.9 ml $O_2 \cdot kg^{-1} \cdot min^{-1}$. From Montoye, H.J., Servais, S.B., & Webster, J.G. (1986). Estimation of energy expenditure from a force platform and an accelerometer. In J. Watkins, T. Reilly, & L. Burwitz (Eds.), *Sports Science*, pp. 375-380. Reproduced with permission of E. and F.N. Spon, an imprint of Chapman and Hall, Ltd., London.

impossible to determine the contribution of the force platform measurements in predicting $\dot{V}O_2$ from their measurements because they did not show the independent contributions of the variables in the multiple regression equations.

Recently, force transducers have been placed under the corners of the floor in a room calorimeter. A very high correlation was reported between indirect calorimetry measures of energy expenditure and force measurements from the floor sensors during walking and stair climbing (Hill & Sun, 1991).

We probably should not expect precise estimates of oxygen consumption from the force plate because considerable energy is expended in internal work and because many activities require isometric and eccentric muscular contractions. De Looze (1992) demonstrated this in repetitive lifting and lowering of loads. The complications discussed by Williams (1985) apply to estimates of energy expenditure from force plates as well as to the photographic methods.

Energy Consumption

Nutritionists and others have estimated energy expenditure by measuring the energy in food consumed, estimates that are accurate only if the individual is in energy balance. Usually this is assumed to be the case if the person's weight has not changed. However, this assumption can be a source of error, if, for example, the weight is unchanged but there is a change in percent body water, body fat, or muscle mass.

Energy intake can be assessed by measuring equal amounts of the food consumed and determining the caloric equivalent of these portions with a "bomb calorimeter." The food is placed in an insulated container, and oxygen is then admitted under high pressure and the contents are ignited. The heat produced is a measure of the energy content of the food. However, an individual is not 100% efficient in absorbing food in the digestive process, so the energy content of excreta should also be determined and subtracted from the energy value of the food ingested to estimate energy utilization.

These are obviously complex laboratory procedures that are not feasible except in controlled circumstances such as a clinical setting. To apply the method in the field, simpler procedures are used. Sometimes food consumption is measured and the energy equivalent determined from food tables, although this involves some error in estimating the energy content of the food. But even this procedure is not practical in many studies and in some clinical and educational circumstances, so the individual may instead be asked to record or recall the food and portion size consumed (Post, 1989). Now additional errors are likely, depending on how accurately and conscientiously the individual records or reports food intake. It is generally accepted that this procedure is not valid among children. Even among adults, the validity is often questionable.

In recent years the results using this method were compared to the results of the doubly labeled water technique (chapter 3). In nine such studies, assuming weight stability, the error in self-report of energy intake was much larger than previously suspected. Quantifying habitual physical activity based on energy intake thus can lead to serious systematic errors in the results. Besides the inaccuracy in self-report information, a problem may arise as a result of the assumption that the energy intake recorded over a short period of 1 to 3 days balances the energy expenditure over the period. This assumption is generally valid only if the period of observation is longer (7 days or more). Therefore, even though energy consumption has been adapted to field conditions, the likelihood of errors in the method is so great that energy consumption outside the clinic or laboratory cannot be considered a valid indicator of daily physical activity.

SUMMARY

The physiological and biomechanical principles underlying movement and energy expenditure have been described. Although these methods are not directly applicable to epidemiologic studies of assessing habitual physical activity, they are important to understanding the techniques for validation of simple practical field methods.

The physiologic methods measuring energy expenditure can be divided into three groups:

- Measurement of energy consumption (food intake). These estimations from energy intake are valid only if there is a state of energy balance.
- Direct measurement of energy expenditure from heat production in a sealed, insulated chamber.
- Indirect measurement of energy expenditure from oxygen consumption in a respiration chamber or using procedures with "closed and open circuit" methods using a hood, small face mask, or nose clip and mouthpiece.

The biomechanical methods measure muscular activity. Displacements and acceleration of whole body or body segments can be registered in two ways:

- Photographs with high-speed camera or video with very elaborate analysis. Automatic registration systems and computerized analysis have been recent improvements.
- Force transducers positioned on the corners of a force plate, a method confined to the laboratory.

The complications with the biomechanical methods are in the use of displacements and acceleration of body and body segments to estimate energy expenditure.

REFERENCES

Atwater, W.O., & Benedict, F.G. (1903). *Experiments on the metabolism of matter and energy in the human body, 1900-1902*. (U.S. Department of Agriculture Office of Experiment Stations, Bulletin 136). Washington, DC: U.S. Government Printing Office.

Atwater, W.O., & Benedict, F.G. (1905). *A respiration calorimeter with appliances for the direct determination of oxygen*. Carnegie Institute of Washington, Publication No. 42.

Barany, T.W., Ismail, A.H., & Manning, K.R. (1965). A force platform for the study of hemiplegic gait. *Journal of the American Physical Therapy Association*, **45,** 693-699.

Brody, S. (1974). *Bioenergetics and growth*. New York: Collier Macmillan. (Original work published 1945)

Brouha, L. (1960). *Physiology in industry*. New York: Pergamon Press.

Carlson, L.D., & Hsieh, A.C.L. (1970). *Control of energy exchange*. New York: Macmillan.

Chapman, C.B., & Mitchell, J.H. (1965). The physiology of exercise. *Scientific American*, **212,** 88-96.

Collins, K.J., Abdel-Rahman, T.A., & Awad El Karim, M.A. (1988). Schistosomiasis: Field studies of energy expenditure in agricultural workers in the Sudan. In K.J. Collins & D.F. Roberts (Eds.), *Capacity for work in the tropics* (pp. 235-247). Cambridge: Cambridge University Press.

Cunningham, D.M., & Brown, G.W. (1952). Two devices for measuring forces acting in the human body during walking. *Proceedings of the Society for Experimental Stress Analysis*, **9,** 75-90.

De Looze, M.P. (1992). *Mechanics and energetics of repetitive lifting*. (GIB Publication No. 4). Thesis, Vrye Universitets, Amsterdam.

Elftman, H. (1938). The measurement of the external force in walking. *Science*, **88,** 152-153.

Fenn, W.O., & Rahn, H. (Eds.) (1964). *Handbook of physiology: Sect. 3. Respiration: Vol. 1*. Washington, DC: American Physiological Society.

Garrow, J.S. (1978). *Energy balance and obesity in man* (2nd ed.). Amsterdam: Elsevier/North Holland Biomedical Press.

Greene, J.H., & Morris, W.H.M. (1958). The force platform: An industrial engineering tool. *Journal of Industrial Engineering*, **9,** 128-132.

Hambraeus, L., Forslund, A., & Sjödin, A. (1991). The use of a suit calorimeter in combination with indirect respiratory calorimetry for studies on the effect of diet and exercise on energy balance in healthy adults [abstract]. *Federation of American Societies for Experimental Biology Journal*, **5,** A1648.

Harrison, M.H., Brown, G.A., & Belyasin, A.J. (1982). The oxylog: An evaluation. *Ergonomics*, **25(a),** 809-820.

Hill, J.O., & Sun, M. (1991). Measurement of the energy cost of activity in humans using a whole-room indirect calorimeter equipped with a mechanical floor [abstract]. *Federation of American Societies for Experimental Biology Journal*, **5,** A1649.

Humphrey, S.J.E., & Wolff, H.S. (1977). The oxylog [abstract]. *Journal of Physiology*, **267,** 120.

Ikegami, Y., Hiruta, S., Ikegami, H., & Miyamura, M. (1988). Development of a telemetry system for measuring oxygen uptake during sports activities. *European Journal of Applied Physiology*, **57,** 622-626.

Insull, W. (1954). *Indirect calorimetry by new techniques: A description and evaluation*. (Report No. 146, Medical Nutrition Laboratory). Denver, CO: U.S. Army, Fitzsimmons Army Hospital.

Ismail, A.H. (1968). Analysis of normal gaits utilizing a special force platform. *Biomechanics I, Proceedings, First International Seminar, 1967, Zurich* (pp. 90-95). Basel, Switzerland: Karger.

Ismail, A.H., Barany, J.W., & Smith, C.B. (1970). Relationship between mechanical force and physiological cost during gait in adult men. In J. Cooper (Ed.), *Proceedings, Biomechanics Symposium, Indiana University* (pp. 99-106). Chicago: Athletic Institute.

Jéquier, E., Acheson, K., & Schutz, Y. (1987). Assessment of energy expenditure and fuel utilization in man. *Annual Review of Nutrition,* **7,** 187-208.

Kinney, J.M. (1980). The application of indirect calorimetry to clinical studies. *Assessment of energy metabolism in health and disease.* Columbus, OH: Ross Laboratories. pp. 42-48.

Kofranyi, E., & Michaelis, H.F. (1940). Ein tragbarer Apparat zur Bestimmung des Gasstoffwechsels [A portable apparatus to determine metabolism]. *Arbeitsphysiologie,* **11,** 148-150.

Kreighbaum, E., & Barthels, K.M. (1990). *Biomechanics* (3rd ed.). New York: Macmillan.

Lauru, L. (1954). The manager. *Management Publications,* **22,** 369.

Lauru, L. (1957). Physiological study of motions. *Advanced Management,* **22,** 17-24.

McNeill, G., Cox, M.D., & Rivers, J.P.W. (1987). The oxylog oxygen consumption meter: A portable device for measurement of energy expenditure. *American Journal of Clinical Nutrition,* **45,** 1415-1419.

Montoye, H.J., Servais, S.B., & Webster, J.G. (1986). Estimation of energy expenditure from a force platform and an accelerometer. In J. Watkins, T. Reilly, & L. Burwitz (Eds.), *Sport Science* (pp. 375-380). London: E. & F.N. Spon.

Montoye, H.J., VanHuss, W.D., Reineke, E.P., & Cockrell, J. (1958). An investigation of the Müller-Franz calorimeter. *Arbeitsphysiologie,* **17,** 28-33.

Müller, E.A., & Franz, H. (1952). Energieverbraughsmessungen bei beruflicher Arbeit mit einer verbesserten Respirationsgasuhr [The measurement of energy consumption during occupational work with an improved respirometer]. *Arbeitsphysiologie,* **14,** 499-504.

Orsini, D., & Passmore, R. (1951). The energy expended carrying loads up and down stairs: Experiments using the Kofranyi-Michaelis calorimeter. *Journal of Physiology,* **115,** 95-100.

Patterson, R.P., & Fisher, S.V. (1979). Energy measurements in ambulation and activities of daily living: Development and evaluation of a portable measuring system [abstract]. *Archives of Physical and Medical Rehabilitation,* **60,** 534-535.

Perkins, J.F. (1964). Historical development of respiratory physiology. In W.O. Fenn & H. Rahn (Eds.) *Handbook of Physiology: Sec. 3. Respiration: Vol. 1,* Washington, DC: American Physiological Society.

Post, G.B. (1989). Nutrition in adolescence: A longitudinal study in dietary patterns from teenager to adult. Thesis, University of Amsterdam, Haarlem, De Vrieseborch, S.O. 16.

Webb, P. (1980). The measurement of energy exchange in man: An analysis. *American Journal of Clinical Nutrition,* **33,** 1299-1310.

Webb., P. (1985). *Human calorimeters.* New York: Praeger.

Webb, P., Annis, J.F., & Troutman, S.J., Jr. (1980). Energy balance in man measured by direct and indirect calorimetry. *American Journal of Clinical Nutrition,* **33,** 1287-1298.

Webb, P., Saris, W.H.M., Schoffelen, P.F.M., van Ingen Schenau, G.J., & ten Hoor, F. (1988). The work of walking: A calorimetric study. *Medicine and Science in Sports and Exercise,* **20,** 331-337.

Weir, J.B. de V. (1949). New methods of calculating metabolic rate with special references to protein metabolism. *Journal of Physiology (London),* **109,** 1-9.

Williams, K.R. (1985). The relationship between mechanical and physiological energy estimates. *Medicine and Science in Sports and Exercise,* **17,** 317-325.

Wolff, H.S. (1958). The integrating pneumotachograph: A new instrument for the measurement of energy expenditures by indirect calorimetry. *Quarterly Journal of Exercise Physiology,* **43,** 270-283.

Zuntz, N., & Leowy, A. (1909). *Lehrbuch der Physiologie bes Menschen* [Textbook of Human Physiology]. Leipzig, Germany: F.C.W. Vogel.

Part II

Field Methods of Assessment

Chapter 3

Doubly Labeled Water

The method of using doubly labeled water (DLW) is a relatively new approach for estimating energy expenditure and is potentially applicable to both laboratory and field studies. The technique's expense, however, makes it impractical for use in large epidemiologic studies or in educational programs. We include it mainly because the body of evidence is growing that we can consider DLW the gold standard for assessing energy expenditure in real-life situations. It has the potential to serve as a criterion for validating more practical approaches.

The principle of the method is simple. A quantity of water with a known concentration of isotopes of hydrogen and oxygen is ingested. The concentration, of course, must be greater than what occurs in nature. In a matter of hours the isotopes distribute themselves in equilibrium with body water. The labeled hydrogen then gradually leaves the body as water (^2H_2O), principally as urine, sweat, and water vapor during respiration. The labeled oxygen also leaves the body as water ($H_2{}^{18}O$) but also as carbon dioxide ($C^{18}O_2$). From the difference in elimination rates of the two isotopes, the production of carbon dioxide can be calculated. Then, by knowing or estimating the respiratory quotient, one can calculate oxygen uptake for the time period.

History of the Method

The term *isotope,* coined by Soddy in 1913, means one of two or more species of the same chemical element that have different atomic weights. The word is derived from two Greek words: *isos,* meaning equal, and *topos,* meaning place, which refers to substances occupying the same place in the periodic table of elements. In 1906 Bolt-Wood observed that two substances that have the same chemical properties need not be physically identical (Encyclopedia Britannica, 1973). Because chemical properties are determined by the positive charge on the nucleus (atomic number) and by the number of extranuclear electrons, isotopes have nuclei with the same number of protons but varying numbers of neutrons. Isotopes can be classified as either stable or unstable, unstable denoting those whose nuclei disintegrate spontaneously—that is, they're radioactive.

In the DLW method one must measure the concentration (i.e., abundance) of stable isotopes of hydrogen (^2H) and oxygen (^{18}O) by separating the isotopes on the basis of very small differences in the masses of their nuclei. This requires a very sensitive isotope ratio mass spectrometer.

In 1929 Giauque and Johnstone identified the isotope ^{18}O, and 4 years later Lewis obtained an enrichment of heavy water—deuterium, or ^2H (Klein et al., 1984; Klein & Klein, 1986). At about that time 40% deuterium-enriched water cost $15 to $20 per gram. By 1950 the cost had decreased to about 20 cents per gram. In 1949 Lifson, Gordon, Visscher, and Nier showed that ^{18}O in expired CO_2 was in equilibrium with ^{18}O in body water. This enabled Lifson, Gordon, and McClintock (1955) to develop the DLW method of estimating energy expenditure and to test the procedure in 15 mice.

Although the cost of deuterium by that time was not great, ^{18}O was expensive, so the method was not practical for humans or large animals. However, with the advent of more sensitive isotope ratio mass spectrometers, the amount of ^{18}O required, and hence the cost, was substantially reduced. Both isotopes are stable, so there is essentially no risk to subjects (Klein & Klein, 1986). In 1982 Schoeller and van Santen evaluated the technique in four healthy adults. There have been other validation studies since then, mostly using the somewhat artificial environment of a respiration chamber. Only a few investigations could be considered field studies of human beings under normal living conditions.

Field Procedures

The abundance of ^2H and ^{18}O in body water can be estimated from blood, urine, saliva, or other body fluids. Deuterium and ^{18}O are naturally occurring isotopes. The abundance of ^2H is approximately 0.014%, which is to say the ratio of ^2H to ^1H is about .00014 to 1. The abundance of ^{18}O averages about 0.204%, or a ratio of ^{18}O to ^{16}O of .00204 to 1. With heavy elements, isotope ratios are essentially constant throughout the world. The natural abundance of light elements like ^2H and ^{18}O, however, varies to a small degree in various parts of the world. It is lower in arctic regions than in the tropics. For this reason it is necessary to analyze a predose sample of body water. The source of drinking water can have an effect on predose abundance, as can the food source. One would not expect much variation in predose values if the subjects do not leave the area or change food habits drastically during the period of observation. Schoeller (1983) reported that the abundance of the isotopes in five subjects varied only 0.19% from week to week.

The dose should be high enough that at the end of the observation period the enrichment of body water above background level would be detectable. Periods of 2 weeks in adults and 6 to 7 days in children would probably provide maximum precision. Schoeller (1983) suggested 5 to 28 days in adults with an optimal dose of 0.12 g/kg Total Body Water (TBW) deuterated water and 0.3 g/kg TBW $H_2{}^{18}O$, and 3 to 14 days for children with slightly higher doses for neonates (0.16 g/kg TBW ^2H_2O and 0.4 g/kg TBW $H_2{}^{18}O$). If the period of observation is too short, the change in concentration of the isotopes is too small for maximal precision. If the period is too long, the isotope enrichment at the end is too low for precise results. It is not necessary for subjects to fast before or after isotope administration (Calazel, Young, Evans, & Roberts, 1991).

Normal protocol for dosing subjects is to take a control urine or saliva sample before administering the initial dose of isotopes orally. The initial enriched samples are taken 5 hours after dosing (or from a second voiding or saliva sample the next morning if dosing is done in the evening). Thereafter, urine or saliva samples are collected daily (multipoint method) or after 7 and 14 days (two-point method), covering two to three half-lives (14-21 days in adults, depending on activity level, or 5 days in infants).

The disappearance rate of the two isotopes, and hence CO_2 production rate, can be calculated in two ways. The multipoint method uses serial (daily) sampling and the assumption of exponential disappearance. The two-point method uses two samples, one at the beginning and one at the end of the observation period. Cole and Coward (1992) showed that the combined accuracy and precision of the multipoint method is better than the two-point method. Seale, Conway, and Canary (1993) reported on a 7-day validation in the respiration chamber using both methods. They found no differences in the validity of both methods against respiratory calorimetry (multipoint—1.55 ± 2.57%; two point—1.59 ± 4.5%). A compromise between the labor-consuming, more precise multipoint method and the two-point method entails using the two-point method over two periods of 7 days each (total measurement period of 2 weeks). Assuming that energy expenditure does not change drastically, a comparison of CO_2 production in Week 1 with Week 2 can reveal possible analytical errors. CO_2 production can be calculated from isotope elimination with the equation (Schoeller et al., 1986)

$$rCO_2 = \frac{K_O \times D_O - K_H \times D_H}{2 \times f_3} - \frac{f_2 - f_1}{2 \times f_3} \times rGf$$

where K_O, D_O, K_H, and D_H are elimination rates and dilution spaces for ^{18}O and ^2H respectively. Factors f_1, f_2, and f_3 are for fractionation of ^2H in water vapor, ^{18}O in water vapor, and ^{18}O in CO_2, respectively. rGf is the rate of istopically fractionated gaseous water loss. Recently Speakman, Nair, and Goran (1993) published a revised equation for calculating CO_2 production that is thought to improve the accuracy of the DLW method.

The measurement of isotopic abundances in gaseous form is accomplished with an isotope ratio mass spectrometer, originally developed by Nier (1940). For using the DLW method with humans, the spectrometer used to analyze hydrogen and carbon dioxide should have a precision of 10 parts per million; this is possible with modern isotope ratio mass spectrometers.

Assumptions and Sources of Error

Although the doubly labeled water method is more accurate than any other technique for measuring energy expenditure, there are sources of error:

- It is assumed that the number of water molecules in the body remains constant during the observation period, so large, unusual changes in body water should be avoided. The effect of errors in this assumption can be estimated by a change in body weight. A 4% change of body weight would result in an error of about 2% in CO_2 production. Nagy (1980) concluded that if the cyclic change in water pool does not exceed 10%, the error in energy expenditure would be insignificant.

- If there is an exchange of ^2H or ^{18}O with nonaqueous body tissue, an error in dilution space results. There is reason to believe that under normal conditions this results in an overestimation of total body water of about 4% from ^2H and about 1% from ^{18}O. Corrections for these differences can be applied. Schoeller and van Santen (1982) reported that in 10 subjects total body water by ^2H was highly correlated ($r = .988$) with total body water by ^{18}O but that the dilution space by ^2H was about 3% higher than that by ^{18}O. Total body water calculated by saliva, urine, or blood serum produced essentially the same results.

- Fractionization due to loss of the isotopes as vapor can result in error. ^1H will evaporate more rapidly from the skin than ^2H. The error from this source is fairly well established. If it is assumed that about 50% of the water is lost as vapor, an error of 10% in the assumption (i.e., 40% or 60% of water was actually lost as vapor) would produce an error in energy expenditure of about 2.5% (Schoeller, 1983).

- The assumption is made that ^2H and ^{18}O are lost only through H_2O and CO_2. If they are lost through other avenues in the same proportion as water loss, no error would result. Some isotopes are lost through urea and feces, which would produce an error of about 2%. Loss of ^{18}O as urinary sulfate appears not to affect energy expenditure measurements (Thirsk, Krause, Preshaw, & Parsons, 1989).

- It is assumed that the turnover rates of isotopic hydrogen and oxygen are constant for the observation period. This is not true for short periods, but using a regression equation fitted to the various measured isotope concentrations would smooth out these changes.

- It is assumed that urinary water or saliva is a representative sample of total body water, and this appears to be correct for ^2H and ^{18}O.

- The assumption is made that isotopic-enriched water and CO_2 do not reenter the body through inspired air or through the skin such that isotope concentration would be affected. This perhaps would be an issue only in a respiration chamber because of the possibility of inhaling $C^{18}O_2$ and $^2H_2{}^{18}O$ due to the relatively small room (~10 cubic meters).

- Diets vary among people to some extent. If the respiratory quotient cannot be determined, it is usually estimated in Western societies to be .85. An error in RQ of .01 would result in an error in energy expenditure of about 1%.

- There is the possibility of small error due to differences in the ambient air temperature and humidity.

- Day-to-day variance in the abundance of isotopes in the local water, air, and food supply generally appears to be small and insignificant, except in a respiration chamber.

Debate continues relative to the details of the methods of analyses, in particular whether the two-point, regression, or integration method should be used in calculating O_2 uptake. It is not appropriate to discuss this controversy and other analytical details here, but the interested reader may consult the consensus report by the IDECG group (Prentice, 1990) and recent papers, such as Cole and Coward (1992); Coward and Prentice (1985); Klein (1984); Schoeller (1984b, 1985); Seale, Miles, and Bodwell (1989); Speakman and Racey (1986); Welle (1990); Westerterp et al. (1988); Wong et al. (1988).

Validation of the Method

The DLW method with humans has never been validated in a field study, and it is likely impossible to do so at present. What would be the criterion? Since the original research of Lifson and colleagues in 1955 in which the validity of the DLW method was studied in mice, there have been many other validity investigations in small animals, birds, insects, and reptiles. These studies have been reviewed elsewhere (Klein, 1984; Nagy, 1980; Roberts, 1989; Schoeller, 1984a); the method appears to be accurate within about ±8%.

It has been possible to study the validity of the DLW method in humans under controlled or hospital conditions. The best possible validation is to monitor CO_2 production and O_2 intake as can be accomplished in a respiration chamber. Less accurate is the comparison in dietary balance studies in which caloric intake is determined. The results of these investigations are summarized in Table 3.1. Different methods and calculations account for some of the variation in these studies. Despite the source of errors previously noted, the validity of the method under controlled conditions appears very satisfactory (within ±5%). However, the error could conceivably be greater in field studies. For example, in a recent field study using the DLW method (Montoye, 1990), 2 of 30 subjects were found to have very low energy expenditures. Both had high intakes of alcohol—in one case it represented more than 20% of energy intake. It is possible that hydrogen was used in the metabolism of alcohol, resulting in a low differential in the elimination rates of ^2H and ^{18}O.

Table 3.1 Summary of Validation Studies of the Measurement of Energy Expenditure by Doubly Labeled Water in Humans

Reference method	Population	Days of observation	% Accuracy (DLW - ref. method)	% Precision (S.D.)	Reference
Caloric intake/balance	4 adults	13	+2.1	±5.6	Schoeller & V. Santen, 1982
Almost continuous $\dot{V}O_2$	5 adults	6	+5.9	±7.6	Schoeller & Webb, 1984
Respiration chamber	1 adult	4.5	−4.6	—	Klein et al., 1984
Respiration chamber	4 adults	—	+1.9	±2.0 (multipoint)	Coward et al., 1984
Respiration chamber	2 adults	3	−2.5	±4.9	Westerterp et al., 1984
Caloric intake/balance	5 adults	14	+3.3	±5.9	Schueller, Kushner, & Jones, 1986
Respiration chamber	6 adults (low dose)	4	+8.0	±7.7	Schoeller et al., 1986
	3 adults (moderate dose)	4	+4.0	±5.0	
Continuous $\dot{V}O_2$	4 infants	5	−1.4	±4.8	Roberts et al., 1986
Continuous $\dot{V}O_2$	9 infants	5	−0.9	±6.2	Jones et al., 1987
Continuous $\dot{V}O_2$	8 infants	6	−0.7	±12.9	Jones et al., 1988
Respiration chamber	5 sedentary adults	6	+1.4	±3.9	Westerterp, Browns, Saris, & ten Hoor, 1988
	4 active adults	3.5	−1.0	±7.0	
Caloric intake/balance	40 infants	—	−1.9 (formula fed)	±10.3	Wang et al., 1989
			+0.2 (breast-fed)	±8.8	
Caloric balance	4 adults	7	−2.58 to 0.32	—	Seale, Rumpler, Conway, & Miles, 1990
Continuous $\dot{V}O_2$	8 preterm infants	5	−4.5	±6.0	Westerterp, Lafeber, Sulkers, & Sauer, 1991
Respiration chamber	4 lean adults	7	+2.9	±0.8	Ravussin, Harper, Rising, & Bogardlus, 1991
	7 obese adults	7	−4.4	±5.2	
Respiration chamber	9 adults	7	+1.55 +1.59	±2.57 (multipoint) ±4.5 (two-point)	Seale, Conway, & Canary, 1993

Another example of a problem with the DLW method under special circumstances was in the Tour de France bicycle race during extreme levels of daily energy expenditure (Westerterp, Saris, van Es, & ten Hoor, 1986; Saris, van Erp, Browns, Westerterp, & ten Hoor, 1989). Comparison of the DLW values with calculated energy expenditure or energy intake for the three 1-week periods showed increasing discrepancy, up to 25%. Schoeller (1983) mentions that special consideration "will be needed in subjects consuming significant amounts of alcohol, subjects with metabolic disorders, or subjects consuming atypical diets" (p. 1004).

Recent Applications

The applications of the doubly labeled water method are diverse. Some examples follow:

- Estimation of energy expenditure in racing the Tour de France (Westerterp et al., 1986) and climbing Mount Everest (Westerterp, Kayser, Brouns, Herry, & Saris, 1992)
- Comparison of energy expenditure in lean and obese subjects (Bandini, Schoeller, & Dietz, 1990; Lichtman et al., 1992; Prentice et al., 1986)
- Changes in energy expenditure in patients before and after surgery (Novick, Nusbaum, & Stein, 1988; Taggart et al., 1991)
- Energy expenditure in subjects in Guatemala (Stein, Johnson, & Greiner, 1988) and The Gambia (Singh et al., 1989)
- Energy expenditure in patients with spina bifida (Bandini, Schoeller, Fukagawa, & Dietz, 1989), anorexia nervosa (Casper, Schoeller, Kushner, Hnilicka, &

Gold, 1991), and cerebral palsy or myelodysplasia (Bandini, Schoeller, Fukagawa, Wykes, & Dietz, 1991) and in burned children (Goran, Peters, Herndon, & Wolfe, 1990)

- Efficiency of energy utilization in women runners and control women (Mulligan, Schoeller, & Calloway, 1989)
- Measurement of energy expenditure under conditions of changing nutrition (Jones et al., 1988; Bandini et al., 1988; Riumallo, Schoeller, Barrera, Gattas, & Vavy, 1989)
- Measurements of energy expenditure in soldiers in training (Forbes-Ewan, Morrissey, Gregg, & Waters, 1989; DeLany, Schoeller, Hoyt, Askey, & Sharp, 1989)
- Measurement of energy expenditure in female endurance athletes and collegiate swimmers (Edwards et al., 1993; Jones & Leitch, 1993)
- Effect of a 5-month endurance training program in sedentary adults (Meijer et al., 1991) and a 1-month training in obese children (Blaak et al., 1992)
- Comparison of energy utilization in breast-fed and formula-fed infants (Butte et al., 1990)
- Evaluation of the heart rate method in children and adults (Livingstone et al., 1990; Schulz, Westerterp, & Bruck, 1989; Emons, Groeneboom, Westerterp, & Saris, 1992)
- Measurement of energy expenditure during pregnancy and lactation (Forsum, Kabir, Sadurskis, & Westerterp, 1992).

Advantages and Disadvantages

Some of the advantages and disadvantages of the doubly labeled water method are outlined in Table 3.2.

Table 3.2 Advantages and Disadvantages of the Doubly Labeled Water Method for Estimating Energy Expenditure

Advantages	Disadvantages
• Validity for estimating energy expenditure in the laboratory appears to be good, accurate to about 1% to 3% and with a precision of about 4% to 7%. Accuracy and precision under the right conditions in the field would probably also be good. • Method is equally applicable to children and adults. • Energy expenditure is measured over a relatively long period (1-3 weeks). • In the measurement of energy expenditure, a measure of body composition also becomes available. • The method is safe and painless and doesn't encumber the subjects during rest or activity. It is also less likely to influence the subject to alter activities than is true of some other methods.	• The cost of ^{18}O is considerable ($400-$600 per subject), which limits the application to relatively small groups. In time, the cost of the isotope may be less. • Analysis of samples is costly and an expensive, sophisticated ratio mass spectrometer as well as the expertise to operate the equipment is required. Studies are being done to improve or simplify the technique (Barrie & Coward, 1985; Wong, Lee, & Klein, 1987). Also, samples can be shipped long distances to centralized laboratories. • In field studies, because carbon dioxide production and not oxygen utilization is being measured, some error is introduced if the respiratory quotient is not known precisely (as is usually the case). • Total energy expenditure over about 4 to 21 days is measured, so no knowledge is obtained for brief periods of peak expenditure. Likewise, no information about specific activities is available.

SUMMARY

Methods for measuring human energy expenditure are either precise but very restrictive—and thus limited to use over a short period of time—or they are less restrictive and usable over long periods but rather imprecise. A relatively new procedure, the so-called doubly labeled water method (DLW), bridges the gap.

The method measures integral CO_2 production for up to 3 weeks from the difference in elimination rates of the stable isotopes deuterium and oxygen-18 from doubly labeled body water after ingestions of a quantity of water enriched with both isotopes.

Validation against the precise and near-continuous respiratory gas exchange method, such as in a respiration chamber, has demonstrated that the method is accurate (1-3%) and has a precision of 4% to 7%, depending on isotope dose, length of elimination period, and frequency of sampling (two-point vs. multipoint). The method is based on a number of assumptions that must be taken into account depending on the application field.

Although there is still some debate on refinements of the kinetic model that may lead to improved accuracy and precision, we can consider the DLW method the gold standard for assessing energy expenditure. The method has several advantages over other techniques: It requires only periodic sampling of body fluids, it is nonrestrictive and ideally suited for use with free-living subjects, and it has the potential to serve as a criterion for validation of more practical approaches. Because the cost of oxygen-18 for one measurement is high ($400-$600) and the analysis of both isotopes requires an expensive and complex isotope ratio mass spectrometer, the method is limited to use with relatively small groups. Recent applications of the method have included a large variety of subjects: athletes (cycling the Tour de France, climbing Mount Everest, running, swimming), soldiers in training, surgical patients, physically disabled subjects, pregnant women, infants, and obese and malnourished children.

REFERENCES

Bandini, L.G., Schoeller, D.A., & Dietz, W.H. (1990). Energy expenditure in obese and non-obese adolescents. *Pediatric Research,* **27,** 198-203.

Bandini, L.G., Schoeller, D.A., Edwards, J., Young, V.R., Oh, S.H., & Dietz, W.H. (1988). Energy expenditure during carbohydrate overfeeding in obese and non-obese adolescents. *American Journal of Physiology,* **256,** E357-E367.

Bandini, L.G., Schoeller, D.A., Fukagawa, N.K., & Dietz, W.H. (1989). Total daily energy expenditure (TDEE) adolescents with spinal bifida (SB) [abstract]. *Federation of American Societies for Experimental Biology Journal,* **3,** A935.

Bandini, L.G., Schoeller, D.A., Fukagawa, N.K., Wykes, L.J., & Dietz, W.H. (1991). Body composition and energy expenditure in adolescents with cerebral palsy or myelodysplasia. *Pediatric Research,* **29,** 70-77.

Barrie, A., & Coward, W.A. (1985). A rapid analytical technique for the determination of energy expenditure by the doubly labeled water method. *Biomedical Mass Spectrometry,* **12,** 535-541.

Blaak, E.E., Westerterp, K.R., Bar-or, O., Woubers, L.J.M. Saris, W.H.M. (1992). Effect of training on total energy expenditure and spontaneous activity in obese boys. *American Journal of Clinical Nutrition,* **55,** 777-782.

Butte, N.F., Wong, W.W., Ferlic, L., O'Brian, M., Smith, E., Klein, P.D., & Garza, C. (1990). Energy expenditure and deposition of breast-fed and formula-fed infants during early infancy. *Pediatric Research,* **28,** 631-640.

Calazel, C.M., Young, V.R., Evans, W.J., & Roberts, S.B. (1991). Effect of experimental protocol on measurement of carbon dioxide production rate using the doubly labeled water method [abstract]. *Federation of American Societies for Experimental Biology Journal,* **5,** A1647.

Casper, R.C., Schoeller, D.A., Kushner, R., Hnilicka, J., & Gold, S.T. (1991). Total daily energy expenditure and activity level in anorexia nervosa. *American Journal of Clinical Nutrition,* **52,** 1143-1150.

Cole, T.J., & Coward, W.A. (1992). Precision and accuracy of doubly labeled water energy expenditure by multipoint and two-point method. *American Journal of Physiology,* **263,** E965-E973.

Coward, W.A., & Prentice, A.M. (1985). Isotope method for the measurement of carbon dioxide production in man. *American Journal of Clinical Nutrition,* **41,** 659-661.

Coward, W.A., Prentice, A.M., Murgatroyd, P.R., Davies, H.L., Cole, T.J., Sawyer, M., Goldberg, G.R., Halliday, D., & MacNamara, J.P. (1984). Measurement of CO_2 production rate in men using ^2H, ^{18}O labeled H_2O: Comparison between calorimeter and isotope values. In H.J.H. van Es (Ed.), *Human energy metabolism: Physical activity and energy expenditure measurements in epidemiological research based on direct and indirect calorimetry* (pp. 126-128), EuroNut Report 5. Den Haag: CIP.

DeLany, J.P., Schoeller, D.A., Hoyt, R.W., Askey, E.W., & Sharp, M.A. (1989). Field use of $D_2{}^{18}O$ to measure energy expenditure of soldiers at different energy intakes. *Journal of Applied Physiology,* **67,** 1922-1929.

Edwards, J.E., Lindeman, A.K., Mikesky, A.E., & Stager, J.M. (1993). Energy balance in highly trained female endurance runners. *Medicine Science in Sports and Exercise,* **25,** 1398-1404.

Emons, H.J.G.,Groeneboom, D.C., Westerterp, K.R., & Saris, W.H.M. (1992). Comparison of heart rate monitoring with indirect calorimetry and doubly labeled water $^2H_2{}^{18}O$ method for the measurement of energy expenditure in children. *European Journal of Applied Physiology,* **65,** 99-103.

Encyclopedia Britannica 14th edition. Isotope. Chicago: Encyclopedia Britannica.

Forbes-Ewan, C.H., Morrissey, L.L., Gregg, G.C., & Waters, D.R. (1989). Use of doubly labeled water technique in soldiers training for jungle warfare. *Journal of Applied Physiology,* **67,** 14-18.

Forsum, E., Kabir, N., Sadurskis, A., & Westerterp, K.R. (1992). Total energy expenditure of healthy Swedish women during pregnancy and lactation. *American Journal of Clinical Nutrition,* **A56,** 334-342.

Goran, M.I., Peters, E.J., Herndon, D.N., & Wolfe, R.R. (1990). Total energy expenditure in burned children using the doubly labeled water technique. *American Journal of Physiology,* **259,** E576-E585.

Jones, P.J.H., Winthrop, A.L., Schoeller, D.A., Sawyer, P.R., Smith, J., Filler, R.M., & Heim, T. (1987). Validation of doubly labeled water for assessing energy expenditure in infants. *Pediatric Research,* **21,** 242-246.

Jones, P.J.H., Winthrop, A.L., Schoeller, D.A., Filler, R.M., Sawyer, P.R., Smith, J., & Heim, T. (1988). Evaluation of doubly labeled water for measuring energy expenditure during changing nutrition. *American Journal of Clinical Nutrition,* **47,** 799-804.

Jones, P.J., & Leitch, C. (1993). Validation of doubly labeled water for measurement of caloric expenditure in collegiate swimmers. *Journal of Applied Physiology,* **74,** 2909-2914.

Klein, P.D. (1984). Letter to the editor. *Human Nutrition: Clinical Nutrition,* **38C,** 479-480.

Klein, P.D., James, W.P.T., Wong, W.W., Irving, C.S., Murgatroyd, P.R., Caberera, M., Dallosso, H.M., Klein, E.R., & Nichols, B.L. (1984). Calorimetric validation of the doubly-labeled water method for determination of energy expenditure in man. *Human Nutrition: Clinical Nutrition,* **38C,** 95-106.

Klein, P.D., & Klein, E.R. (1986). Stable isotopes: Origins and safety. *Journal of Clinical Pharmacology,* **26,** 378-382.

Lichtman, S.W., Pisarswa, K., Berman, E.R., Pestone, M., Dowling, H., Offenbacher, E., Weisel, H., Heshka, S., Matthews, D.E., & Heymsfield, D. (1992). Discrepancy between self-reported and actual calorie intake and exercise in obese subjects. *New England Journal of Medicine,* **327,** 1893-1896.

Lifson, N., Gordon, G.B., & McClintock, R. (1955). Measurement of total carbon dioxide production by means of D_2O^{18}. *Journal of Applied Physiology,* **7,** 704-710.

Lifson, N., Gordon, G.B., Visscher, M.B., & Nier, A.O. (1949) The fate of utilized molecular oxygen and the source of the oxygen of respiratory carbon dioxide studied with the aid of heavy oxygen. *Journal of Biological Chemistry,* **180,** 803-811.

Livingston, M.B., Prentice, A.M., Coward, W.A., Ceesay, S.M., Strain, J.J., McKenna, P.G., Nevin, G.B., Barker, M.E., & Hickey, R.J. (1990). Simultaneous measurement of free-living expenditure by doubly labeled water method and heart-rate monitoring. *American Journal of Clinical Nutrition,* **52,** 59-65.

Meijer, G.A.L., Janssen, G.M.E., Westerterp, K.R., Verhoeven, F., Saris, W.H.M., & ten Hoor, F. (1991). The effect of a 5-month endurance-training programme on physical activity: Evidence for a sex-difference in the metabolic response to exercise. *European Journal of Applied Physiology,* **62,** 11-17.

Montoye, H.J. (1990). Validation of methods of measuring physical activity, final report for grant 5RO1HL37561, National Institute of Health, p. 22.

Mulligan, K., Schoeller, D.H., & Calloway, D.H. (1989). Energy balance in women runners. *Federation of American Societies for Experimental Biology Journal,* **3,** A1286.

Nagy, K.A. (1980). CO_2 production in animals: Analysis of potential errors in the doubly labeled water method. *American Journal of Physiology,* **238,** R466-R473.

Nier, A.O. (1940). A mass spectrometer for routine isotope abundance measurements. *Review of Scientific Instruments,* **11,** 212-216.

Novick, W.M., Nusbaum, M., & Stein, T.P. (1988). The energy costs of surgery as measured by the doubly labeled water ($^2H_2{}^{18}O$) method. *Surgery,* **103,** 99-106.

Prentice, A.M. (Ed.) (1990). The doubly-labeled water method for measuring energy expenditure, technical recommendations for use in humans. Vienna: Atomic Energy Agency.

Prentice, A.M., Black, A.E., Coward, W.A., Davies, H.L., Goldberg, G.R., Murgatroyd, P.R., Ashford, J., Sawyer, M., & Whitehead, R.G. (1986). High levels of energy expenditure in obese women. *British Medical Journal,* **292,** 983-987.

Ravussin, E., Harper, I.T., Rising, R., & Bogardus, C. (1991). Energy expenditure by doubly labeled water: Validation in lean and obese subjects. *American Journal of Physiology,* **261,** E402-E409.

Riumallo, J.A., Schoeller, D., Barrera, G., Gattas, V., & Vavy, R. (1989). Energy expenditure in underweight free-living adults: Impact of energy supplementation as determined by doubly labeled water and indirect calorimetry. *American Journal of Clinical Nutrition,* **49,** 239-246.

Roberts, S.B. (1989). Use of doubly labeled water method for measurement of energy expenditure, total body water, water intake, and metabolizable energy intake in humans and small animals. *Canadian Journal of Physiology and Pharmacology,* **67,** 1190-1198.

Roberts, S.B., Coward, W.A., Schlingenseipen, K.H., Mohris, V., & Lucas, A. (1986). Comparison of the doubly labeled water ($^2H_2{}^{18}O$) method with indirect calorimetry and a nutrient-balance study for simultaneous determination of energy expenditure, water intake, and metabolizable energy intake in pre-term infants. *American Journal of Clinical Nutrition,* **44,** 315-322.

Saris, W.H.M., van Erp, M.A., Browns, B.F., Westerterp, K.R., & ten Hoor, F. (1989). Study on food intake and energy expenditure during extreme sustained exercise. *International Journal of Sports Medicine,* **10,** 510-531.

Schoeller, D.A. (1983). Energy expenditure from doubly labeled water: Some fundamental considerations in humans. *American Journal of Clinical Nutrition,* **38,** 999-1005.

Schoeller, D.A. (1984a). Letter to the editor. *Human Nutrition: Clinical Nutrition,* **38C,** 479-480.

Schoeller, D.A. (1984b). Use of two point sampling for the doubly labeled water method. *Human Nutrition: Clinical Nutrition,* **38C,** 477-479.

Schoeller, D.A. (1985). Reply to letter from Coward & Prentice. *American Journal of Clinical Nutrition,* **41,** 661-662.

Schoeller, D.A., Kushner, R.F., & Jones, P.H. (1986). Validation of doubly labeled water for measuring energy expenditure during parenteral nutrition. *American Journal of Clinical Nutrition,* **44,** 291-298.

Schoeller, D.A., Ravussin, E., Schutz, Y., Acheson, K.J., Baertschi, P., & Jéquier, E. (1986). Energy expenditure by doubly labeled water: Validation in humans and proposed calculation. *American Journal of Physiology,* **250,** R823-R830.

Schoeller, D.A., & van Santen, E. (1982). Measurement of energy expenditure in humans by doubly labeled water method. *Journal of Applied Physiology: Respiration, Environment, and Exercise Physiology,* **53,** 955-959.

Schoeller, D.A., & Webb, P. (1984). Five-day comparison of the doubly labeled water method with respiratory gas exchange. *American Journal of Clinical Nutrition,* **40,** 153-158.

Schulz, S., Westerterp, K.R., & Bruck, K. (1989). Comparison of energy expenditure by doubly labeled water technique with energy intake, heart rate and activity recording in man. *American Journal of Clinical Nutrition,* **49,** 1148-1154.

Seale, J.L., Conway, J.M., & Canary, J.J. (1993). Seven day validation of doubly labeled water method using indirect room calorimetry. *Journal of Applied Physiology,* **74,** 402-409.

Seale, J.,Miles, C., & Bodwell, C.E. (1989). Sensitivity of methods for calculating energy expenditure by use of doubly labeled water. *Journal of Applied Physiology,* **66,** 644-653.

Seale, J.L., Rumpler, W.V., Conway, J.M., & Miles, C.W. (1990). Comparison of doubly labeled water, intake-balance, and direct- and indirect-calorimetry methods for measuring energy expenditure in adult men. *American Journal of Clinical Nutrition,* **52,** 66-71.

Singh, J., Prentice, A.M., Diaz, E., Coward, W.A., Ashford, J., Sawyer, M., & Whitehead, R.G. (1989). Energy expenditure of Gambian women during peak agricultural activity measured by the doubly-labeled water method. *British Journal of Nutrition,* **62,** 315-329.

Speakman, J.R., Nair, K.S., & Goran, M.I. (1993). Revised equations for calculating CO_2 production from doubly labeled water in humans. *American Journal of Physiology,* **264,** E912-E917.

Speakman, J.R., & Racey, P.A. (1986). Measurement of CO_2 production by the doubly labeled water technique. *Journal of Applied Physiology,* **61,** 1200-1202.

Stein, T.P., Johnson, F.E., & Greiner, L. (1988). Energy expenditure and socioeconomic status in Guatemala as measured by the doubly labeled water method. *American Journal of Clinical Nutrition,* **7,** 196-200.

Taggart, D.P., McMillan, D.C., Preston, T., Richardson, R., Burns, H.J.G., & Wheatley, D.J. (1991). Effects of cardiac surgery and intraoperative hypothermia on energy expenditure as measured by doubly labeled water. *British Journal of Surgery,* **78,** 237-241.

Thirsk, J.E., Krause, H.R., Preshaw, R.M., & Parsons, H.G. (1989). Is oxygen-18 lost as urinary sulfate and does this affect the doubly labeled water method of measuring energy expenditure? [abstract]. *Federation of American Societies for Experimental Biology Journal,* **3,** 727.

Welle, S. (1990). Two point versus multipoint sample collection for the analysis of energy expenditure by use of the doubly labeled water method. *American Journal of Clinical Nutrition,* **52,** 1134-1138.

Westerterp, K.R., Browns, F., Saris, W.H.M., & ten Hoor, F. (1988). Comparison of doubly labeled water with respirometry at low- and high-activity levles. *Journal of Applied Physiology,* **65,** 53-56.

Westerterp, K.R., De Boer, J.O., Saris, W.H.M., Schoffelen, P.F.M., & ten Hoor, F. (1984). Measurement of energy expenditure using doubly labeled water. *International Journal of Sports Medicine,* **5,** 74-75.

Westerterp, K.R., Kayser, B., [Brouns, F.], Herry, J.P., & Saris, W.H.M. (1992). Energy expenditure climbing the Mt. Everest. *Journal of Applied Physiology,* **73,** 1815-1819.

Westerterp, K.R., Lafeber, H.N., Sulkers, E.J., & Sauer, P.J.J. (1991). Comparison of short term indirect calorimetry and doubly labeled water method for the measurement of energy expenditure in infants. *Biology of the Neonate,* **60,** 75-82.

Westerterp, K.R., Saris, W.H.M., van Es, M., & ten Hoor, F. (1986). Use of the doubly labeled water technique in humans during heavy sustained exercise. *Journal of Applied Physiology,* **61,** 2162-2167.

Wong, W., Butte, N., Patterson, B., Clarke, L., Garza, C., & Klein, B. (1989). Validation of the $^2H_2{}^{18}O$ method for estimation of energy expenditure in infants [abstract]. *Federation of American Societies for Experimental Biology Journal,* **3,** A934.

Wong, W.W., Cochran, W.J., Klish, W.J., Smith, E.O., Lee, L.S., & Klein, P.D. (1988). In vivo isotope-fractionation factors and the measurement of deuterium- and oxygen-18 dilution spaces from plasma, urine, saliva, respiratory water vapor, and carbon dioxide. *American Journal of Clinical Nutrition,* **47,** 1-6.

Wong, W.W., Lee, L.S., & Klein, P.D. (1987). Deuterium and oxygen-18 measurement on microliter samples of urine, plasma, saliva and human milk. *American Journal of Clinical Nutrition,* **45,** 905-913.

Chapter 4

BEHAVIORAL OBSERVATION AND TIME/MOTION ANALYSES

One of the earliest methods employed to assess physical activity entailed an observer's recording observations while watching a subject. This technique has frequently been used to view workers on the job in studying efficiency or fatigue, to study preschool and schoolchildren when other methods were not feasible, and to evaluate other techniques of estimating physical activity or energy expenditure.

For want of a suitable criterion, observations for estimating physical activity and energy expenditure have never been adequately validated. However, face validity seems good when observations are carefully made and recorded. The cost of observation is often prohibitive, particularly if it is necessary to employ one observer for each subject. Observers may make note of various data: behavioral information, types of activities, frequency of performance, and time per activity. Estimating the energy cost of each activity and multiplying this by the time allocation provides an estimate of total energy expenditure. Appendix C lists the estimated energy costs of various activities.

One drawback to observation is that subjects may alter their usual activity when they know it is being observed. Observing and recording also is tedious, and accuracy probably decreases as the observation period lengthens. And, of course, errors are introduced when using values such as those in Appendix C to estimate energy expenditure. The observation method generally has been used in small studies or as a standard to which other techniques might be compared.

Simple Observation Methods

For data gathered by observers to be reduced to a usable form, observations and recordings must be systematized. Some very simple methods have been used. Wade and Ellis (1971) classified activities of kindergartners into four categories: sitting quietly, sitting accompanied by small movements, walking, and running. In another study (Thorland & Gilliam, 1981), parents rated their young children's activities on a 6-point scale, ranging from 1 to more than 10 METs. Saris and Binkhorst (1977) used four categories of activity, each at two levels of intensity, with kindergarten children. Three categories of strenuousness of activities were used with 8-year-old children by Ku, Shapiro, Crawford, and Huenemann (1981). Torún (1984) recorded 40 different activities in preschool children. At the extreme, Fales (1938) listed 651 categories of activity.

Usually a form is prepared for recording activities. Several forms are illustrated in Figures 4.1 through 4.4 (Torún, 1984; Wallace, McKenzie, & Nader, 1985; Baranowski et al., 1984; Hovell, Bursick, Sharkey, & McClure, 1978). The original references should be consulted for details on how these forms are used. A system of observing and coding the strenuousness of activities in a physical education class was developed by McKenzie, Sallis, and Nader (1991). The system, called SOFIT, could be adapted for other uses. Table 4.1, modified from McKenzie, Sallis, Patterson, and Elder (1991), summarizes characteristics, including validity, of many observational studies of children.

Complex Observation Methods

Various techniques and equipment have been used in an effort to improve reliability and provide permanent records of activities. Movie cameras and video recorders create records that may be viewed as many times as desired by any number of observers. This repeatable record does not reduce the cost of the procedure but probably improves objectivity. In the classic study by Bullen, Reed, and Mayer (1964), a 16-mm motion picture camera with a wide-angle lens was used to photograph a number of children simultaneously (the so-called multimovement registration). The camera, mounted on a tripod, panned the group for 3 s every 7 min (volleyball), every 10 or 6 min (swimming), and every 4 min (tennis), providing 27,211 observations of the children. When the film was viewed, the activities and the intensity with which they were performed were recorded and estimates of caloric expenditure made. Time-lapse cinematography has also been used, particularly in industry; activity can be recorded at a slow rate (1 frame per second or slower) and then projected later at normal speed (Edholm, 1966). Thus, hours of observation can be examined in a few minutes. This technique has been used primarily to study efficiency in factory work, but the system could be adapted for observing other activities.

Some years ago Wolff (1959) described a system whereby an observer carrying a walkie-talkie radio set could relay information to a central station, where pertinent information was recorded. The system was applied in a study by Edholm et al. (1970). Wolff (1959) also described an observation system using punch cards. Even earlier, Welford (1952) described a serial event timer and recorder (SETAR) that utilizes a teleprinter to record events. The unit, about 4 ft high, is not readily used in the field. Portable computers have rendered both the Wolff and Welford apparatuses obsolete for most applications.

TIME	LYING DOWN		SITTING		STANDING		WALKING		RUNNING		PLAYING		NUMBER OF OBSERVATIONS
	asleep	awake	still	active	still	active	slowly	rapidly	slowly	rapidly	quietly	actively	
8:00 8:10		✓✓ 6.7*									✓ 3.3		3 (3.3' each)
8:10 8:20						✓ 2.5	✓✓ 5		✓ 2.5				4 (2.5' each)
8:20 8:30				✓ 10									1 (10')
8:30 8:40			✓ 2	✓ 2		✓✓ 4				✓ 2			5 (2' each)
: : : : : : : :													

*Minutes spent in each activity = (10 minutes/total number of observations) x number of times the activity was observed.

Figure 4.1 A simplified form for recording activities of a child in time and motion studies. From Torún, B. (1984). Physiological measurements of physical activity among children under free-living conditions. In E. Pollitt & P. Amante (Eds.), *Energy Intake and Activity*, pp. 159-184. Copyright © 1984 by Alan R. Liss. Reprinted by permission of Wiley-Liss, a division of John Wiley and Sons, Inc.

Activity Comparison

Name Jan Doe Day Thurs. No. 7

Activity: Counselor	Activity: Camper	Time: Counselor	Time: Camper
Sleep		8.25 hrs.	9.0 hrs.
Easy			
Moderate			
Warm-ups	Warm-ups	15 min	15 min
Weight Training		30	
New Games		15	
	Basketball		30
Hard			
Aerobics		45	
Swimming	Swimming	45	45
Tennis	Tennis	45	45
Bike		15	
	Basketball		45
	Dance		60
Very Hard			
Dance		60	
	Aerobics		45
	Games		45

Counselor Activity Record

Name Jan Doe Day-Date Thursday, 14 July 1983

Counselor Jim Camp

Time	Activity	Time	Activity
6:00 am	Sleep	6:00 pm	Sat around
6:15	Woke up	6:15	↓
6:30	Dressed	6:30	
6:45	Warm-up Exercises	6:45	
7:00	Walked .6 mi	7:00	All Camp Activity
7:15	Breakfast	7:15	danced the entile
7:30	↓	7:30	time ↓
7:45	Walked .6 mi	7:45	
8:00	Aerobic Class-Walking	8:00	New Games-running
8:15	↓	8:15	ate snack
8:30		8:30	↓
8:45		8:45	Showered
9:00	Swimming Class	9:00	↓
9:15	↓	9:15	
9:30		9:30	Bed
9:45	Changed Clothes	9:45	↓
10:00	Walk	10:00	
10:15	Tennis Class	10:15	
10:30	↓	10:30	
10:45		10:45	
11:00	Walked	11:00	Sleep
11:15	Weight training	11:15	↓
11:30	↓	11:30	
11:45	Walked .6 mi	11:45	
12:00	Lunch	12:00	
12:15	↓	12:15	
12:30		12:30	
12:45	Walked .6 mi	12:45	
1:00	Sat around	1:00	
1:15	↓		
1:30			
1:45			
2:00	Behavior Mod		
2:15	↓		
2:30	Nut		

Figure 4.2 A form for a counselor to record physical activities of a child. This form is reprinted with permission from Wallace, J.P., McKenzie, T.L., & Nader, P.R. (1985). Observed versus recalled exercise behavior: A validation of a seven-day exercise recall for boys 11 to 13 years old. *Research Quarterly for Exercise and Sport*, **56**, 161-165. The *Research Quarterly for Exercise and Sport* is a publication of the American Alliance for Health, Physical Education, Recreation and Dance, 1900 Association Drive, Reston, VA 22091, U.S.A.

OBSERVATION FORM

OBSERVER ____________________

Subject I.D. ____________________ Hour ____________________

			2	4	6	8	10	12	14	16	18	20	22	24	26	28	30	32	34	36	38	40	42	44	46	48	50	52	54	56	58	60
Physical location:																																
Home:	Personal bedroom																															
	Bathroom																															
	Kitchen																															
	Living room																															
	Dining room																															
	Other:																															
School:	Lunchroom/Cafeteria																															
	Gymnasium																															
	Playfield																															
	Other:																															
Outside:	Around home																															
	Around school																															
	Other:																															
	Missed observation																															
Social environment																																
	Alone																															
	Mother																															
	Father																															
	Sister																															
	Brother																															
	Peer																															
	Observer-on																															
	Observer-off																															
	Other:																															
Groups:	Family (parents & sibs)																															
	Parents																															
	Siblings																															
	Peers																															
	Others:																															
	Missed observations																															
	Stationary																															
Activity	Team Sport:	slow																														
		fast																														
	Walking:	slow																														
		fast																														
	Running/Jogging																															
	Bicycling:	slow																														
		fast																														
	Other: (1)	slow																														
		fast																														
	Other: (2)	slow																														
		fast																														
	Sleeping																															
	Missed observation																															
	HR																															
	School bus/No observation																															

Figure 4.3 A form for recording a child's activities during 1 hr. This form is reprinted with permission from Baranowski, T., Dworkin, R.J., Cieslik, C.J., Hooks, P., Clearman, D.R., Ray, L., Dunn, J.K., & Nader, P.R. (1984). Reliability and validity of self-report of aerobic activity: Family health project. *Research Quarterly for Exercise and Sport*, **55**, 309-317. The *Research Quarterly for Exercise and Sport* is a publication of the American Alliance for Health, Physical Education, Recreation and Dance, 1900 Association Drive, Reston, VA 22091, U.S.A.

Child's Number ______________ Male ________ Female ________

School ______________________________

Time of Observation ______________________________

Day ______________________ Date ________

Seconds	5	10	15	20	25	30	35	40	45	50
Upper	A B C	A B C	A B C	A B C	A B C	A B C	A B C	A B C	A B C	A B C
Lower	A B C	A B C	A B C	A B C	A B C	A B C	A B C	A B C	A B C	A B C
Upper	A B C	A B C	A B C	A B C	A B C	A B C	A B C	A B C	A B C	A B C
Lower	A B C	A B C	A B C	A B C	A B C	A B C	A B C	A B C	A B C	A B C

Figure 4.4 A form used to record the activity of a child's upper and lower body for two 50-s periods. A = little or no activity; B = moderately intense activity; C = extensive vigorous activity. This form is reprinted with permission from Hovell, M.F., Bursick, J.H., Sharkey, R., & McClure, J. (1978). An evaluation of elementary students' voluntary physical activity during recess. *Research Quarterly for Exercise and Sport*, **49**, 460-474. The *Research Quarterly for Exercise and Sport* is a publication of the American Alliance for Health, Physical Education, Recreation and Dance, 1900 Association Drive, Reston, VA 22091, U.S.A.

An inexpensive event-counting digital recording system that is compatible with automated data analysis equipment has been developed by McPartland, Foster, and Kupfer (1976). The coded activities may be digitally logged on inexpensive magnetic tape cassettes. This system is compatible with automated data analysis equipment and is adaptable for many observational studies. At about the same time, a group at the University of Wisconsin developed a computer-compatible system for observing animal behavior (Goldfoot, Essock-Vitale, Asa, Thornton, & Leshner, 1978; Stephenson & Roberts, 1977; Stephenson, Smith, & Roberts, 1975). Because it is battery operated, it is suitable for field use for animals and humans. The incidence, sequence, and duration of various coded activities are entered on a keyboard and stored on magnetic tape, and high-speed analysis is subsequently done by computer.

A computer-registered direct observation system known as *task recording and analysis on computer* (TRAC) was developed recently (Vanderbeek, van Gaalen, & Frings-Dresen, 1992). It is a flexible system employing a pocket calculator. Activity and posture

Table 4.1 Characteristics of Selected Instruments Designed for Observing Children's Physical Activity

Observation strategy	Activity categories	Validation	Test site	Subjects	Reliability	References
Momentary time sampling: 5-s intervals	3	None	Recess	3rd- to 6th-grade students	I-I (88%-94%)	Hovell, Bursick, Sharkey, & McClure, 1978
Momentary time sampling: 15-s intervals	5	Heart rates	Free play in gym	Obese girls, ages 5-8	I-I (86%-99%)	Epstein, McGowan, & Woodall, 1984
Partial time sampling: 1-min intervals	4	Heart rates	Physical education classes	3rd- to 5th-grade students	I-I (95%-98%)	O'Hara, Baranowski, Simons-Morton, Wilson, & Parcel, 1989
Momentary time sampling: 10-s observation/ recording intervals	5	Heart rates	Physical education classes	3rd- to 5th-grade students	I-I (92%)	McKenzie, Sallis, & Nader, 1991
Partial interval recording: 10-s observation/ recording intervals	8	None: correlated with LSI readings[a]	Home	Children, ages 20-48 months	I-I (91%-98%) Kappa (.90)	Klesges et al., 1984
Partial time sampling: five 1-min intervals	5	Heart rates, VO_2	Diverse	Children, ages 3-6 years	I-I (84.1%)	Puhl, Greaves, Hoyt, & Baranowski, 1990
Momentary time sampling: 1-min intervals	5	Heart rates	Diverse	Children, ages 4-9 years	Kappa (.91)	McKenzie, Sallis, Patterson, & Elder, 1991
Momentary time sampling: 10-s observation/ recording intervals	4	None	Home	Children, ages 3-6 years	Field scores, NA; videotape Kappa (.91)	Klesges, Eck, Hanson, Haddock, & Klesges, 1990
Time sampling	6	CHD risk factors[b]	Diverse	55 boys	$r = .84$	Thorland & Gilliam, 1981
Momentary time sampling: 2-min intervals	4	Pedometer and actometer	School	11 children, ages 5-6 years	$r = .93$ $r = .95$	Saris & Binkhorst, 1977

[a]LSI = Large-Scale Integrated Motor Activity Monitor.
[b]CHD = Coronary heart disease.

Modified from T. McKenzie, J. Sallis, T. Patterson, and J. Elder, 1991, Observational measures of children's physical activity. *Journal of School Health*, **61**, 224-227. Reproduced with permission.

variables, multimovement observations, and real time can all be recorded and combined with heart rate data.

Reliability

Observations of physical activity are often described without reporting the reproducibility (reliability) of the method. However, this is not always the case. Klesges et al. (1984) described the Fargo Activity Timesampling Survey (FATS) for observing physical activity in children. Eight activities, from very sedentary to strenuous, are listed, with three levels of intensity for each activity. The child is observed for 10 s and the activity recorded in the next 10 s, alternating in this way for the period of observation. The percentage of time in each activity and at each intensity can be tabulated. A composite index taking into consideration the activity, intensity, and time duration can be calculated. Observers are trained until an interobserver, frame-by-frame agreement of about 90% is reached (that is, agreements ÷ [agreements + disagreements] × 100).

In observations of two children, 22 and 20 months of age, agreement varied from 90% to 96% between two observers. Among 14 children ages 24 to 48 months, the two observers again agreed 91% to 98% of the time for the various activity components during a 2-hr observation period. A month later, the children were observed again. The test-retest correlation coefficient for the composite index was .59. In another study, Klesges, Woolfrey, and Vollmer (1985), studying 30 children, compared this same 10-s recording system to a continuous activity recording by a second observer who used a hand-held computer device (Datamyte 1002). For the composite activity rating, the two methods agreed quite well ($r = .90$). All things considered, the noncomputer, 10-s time-sampling method appeared to be the one of choice.

Fales (1938) reported a high correlation (.98) between two observers recording 34 five-min intervals of a child's activity. The activity, intensity (vigorousness), and time of each activity were used to calculate a score for the 5-min period. Using the form illustrated in Figure 4.3, Baranowski et al. (1984) reported interobserver agreement in recording data on children in the third to sixth grade (about 8 to 11 years of age). Hovell et al. (1978) also reported interobserver agreement of about 89% using their form (Figure 4.4) with children of the same ages.

Observations have been compared to physical activity assessment by other methods; these comparisons will be summarized in subsequent chapters.

SUMMARY

Assessing physical activity by observation works particularly well with small children, when most other assessment methods are unsuitable. But observation is time-consuming and expensive and thus not suitable for use in even moderately large groups. Also, observations are confined to relatively short periods and may not reflect habitual physical activity. With training, observers can be quite accurate. Various forms are available to make recording more efficient. Also, devices are available, some of them computer compatible, that facilitate the observation approach of assessing physical activity.

REFERENCES

Baranowski, T., Dworkin, R.J., Cieslik, C.J., Hooks, P., Clearman, D.R., Ray, L., Dunn, J.K., & Nader, P.R. (1984). Reliability and validity of self report of aerobic activity: Family health project. *Research Quarterly for Exercise and Sport,* **55,** 309-317.

Bullen, B.A., Reed, R.B., & Mayer, J. (1964). Physical activity of obese and non-obese adolescent girls appraised by motion picture sampling. *American Journal of Clinical Nutrition,* **14,** 211-223.

Edholm, O.G. (1966). The assessment of habitual activity. In K. Evang & K.L. Andersen (Eds.), *Physical Activity in Health and Disease,* (pp. 187-197). Oslo: Oslo University Press.

Edholm, O.G., Adam, J.M., Healy, M.J.R., Wolff, H.S., Goldsmith, R., & Best, T.W. (1970). Food intake and energy expenditure of army recruits. *British Journal of Nutrition,* **24,** 1091-1107.

Epstein, L., McGowan, C., & Woodall, K. (1984). A behavioral observation system for free play activity in young overweight female children. *Research Quarterly for Exercise and Sports,* **55,** 180-183.

Fales, E. (1938). A rating scale of the vigorousness of play activities of preschool children. *Child Development,* **8,** 15-46.

Goldfoot, D.A., Essock-Vitale, S.M., Asa, C.S., Thornton, J.E., & Leshner, A.I. (1978). Anosmia in male rhesus monkeys does not alter copulatory activity with cycling females. *Science,* **199,** 1095-1096.

Hovell, M.F., Bursick, J.H., Sharkey, R., & McClure, J. (1978). An evaluation of elementary students' voluntary physical activity during recess. *Research Quarterly for Exercise and Sport,* **49,** 460-474.

Klesges, R.C., Coats, T., Modenhauer-Klesges, L., Holzer, B., Gustavson, J., & Barnes, J. (1984). The FATS: An observational system for assessing physical activity in children and associated parent behavior. *Behavioral Assessment,* **6,** 333-345.

Klesges, R.C., Woolfrey, J., & Vollmer, J. (1985). An evaluation of the reliability of time sampling versus continuous observation data collection. *Journal of Behavior Therapeutics and Experimental Psychiatry,* **16,** 303-307.

Klesges, R., Eck, L, Hanson, C., Haddock, K., & Klesges, L. (1990) Effects of obesity, social interactions, and physical environment on physical activity in preschoolers, *Health Psychology,* **9,** 435-449.

Ku, L.C., Shapiro, L.R., Crawford, P.B., & Huenemann, R.L. (1981). Body composition and physical activity in 8-year-old children. *American Journal of Clinical Nutrition,* **34,** 2770-2775.

McKenzie, T., Sallis, J., Patterson, T., & Elder, J. (1991). BEACHES: An observational system for assessing children's eating and physical activity behaviors and associated events. *Journal of Applied Behavioral Analysis,* **24,** 1.

McKenzie, T., Sallis, J., & Nader, P. (1991). SOFIT: System for observing fitness instruction time. *Journal of Teaching Physical Education,* **11,** 195-205.

McKenzie, T.L. (1991). Observational measures of children's physical activity. *Journal of School Health,* **61,** 224-227.

McPartland, R.J., Foster, F.G., & Kupfer, D.J. (1976). A computer-compatible multichannel event counting and digital recording system. *Behavioral Research Methods and Instrumentation,* **8,** 299-301.

O'Hara, N., Baranowski, T., Simons-Morton, B., Wilson, S., & Parcel, G. (1989). Validity of the observation of children's physical activity. *Research Quarterly of Exercise and Sports,* **60,** 41-47.

Passmore, R., Thomson, J.G., & Warnock, G.M. (1952). Balance sheet of the estimation of energy intake and energy expenditure as measured by indirect calorimetry. *British Journal of Nutrition,* **6,** 253-264.

Puhl, J., Greaves, K., Hoyt, M., & Baranowski, T. (1990). Children's activity rating scale (CARS): Description and evaluation. *Research Quarterly for Exercise and Sports,* **61,** 26-36.

Saris, W.H.M., & Binkhorst, R.A. (1977). The use of pedometer and actometer in studying daily physical activity in man. Part II: Validity of pedometer and actometer measuring daily physical activity. *European Journal of Applied Physiology,* **37,** 229-237.

Stephenson, G.R., Smith, D.P.S., & Roberts, T.W. (1975). The SSR system: An open format event recording system with computerized transcription. *Behavior Research Methods and Instrumentation,* **7,** 497-515.

Stephenson, G.R., & Roberts, T.W. (1977). The SSR system 7: A general recording system with computerized transcription. *Behavior Research Methods and Instrumentation,* **9,** 434-441.

Thorland, W.G., & Gilliam, T.B. (1981). Comparison of serum lipids between habitually high and low active pre-adolescent males. *Medicine and Science in Sports and Exercise,* **13,** 316-321.

Torún, B. (1984). Physiological measurements of physical activity among children under free-living conditions. In E. Pollitt & P. Amante (Eds.), *Energy intake and activity,* (pp. 159-184). New York: Alan R. Liss.

Vanderbeek, A.J., van Gaalen, L.D., & Frings-Dresen, M.H.W. (1992). Working postures and activities of lorry drivers—A reliability study of on-site observation and recording on a pocket computer. *Applied Ergonomics,* **23,** 331-336.

Wade, M.G., & Ellis, M.J. (1971). Measurement of free-range activity in children as modified by social and environmental complexity. *American Journal of Clinical Nutrition,* **24,** 1457-1460.

Wallace, J.P., McKenzie, T.L., & Nader, P.R. (1985). Observed versus recalled exercise behavior: A validation of a seven day exercise recall for boys 11 to 13 years old. *Research Quarterly for Exercise and Sport,* **56,** 161-165.

Welford, N.T. (1952). An electronic digital recording machine: The SETAR. *Journal of Scientific Instruments,* **29,** 1-4.

Wolff, H.S. (1959). Modern techniques for time and motion study in physiological research. *Ergonomics,* **2,** 354-362.

Chapter 5

THE DIARY METHOD

The diary method of assessing habitual activity consists of logging one's own activities periodically. Recording has been done as often as every minute (Riumallo, Schoeller, Barrera, Gattas, & Vauy, 1989) and as infrequently as every 4 hr (LaPorte, 1979). Very detailed records can be maintained of specific activities (Gorsky & Calloway, 1983), or the individual may note activities only in gross categories (for example, moderate, heavy, or very heavy; Durnin, 1967).

A prepared form may be used to facilitate maintaining the diary. Figures 5.1 through 5.4 show sample forms from Edholm (1966); Bouchard et al. (1983); Durnin and Passmore (1967); and Andersen, Masironi, Rutenfranz, and Seliger (1978). In a diary form like the one from Edholm (Figure 5.1), the subject must fill in each minute (1,440 min during 24 hr) to insure that no minutes are missed. Recording a change in activity (Figures 5.3 and 5.4) is simpler than recording each activity minute by minute. Using symbols as a kind of shorthand also eases the burden of making diary entries (Andersen et al., 1978; Collins & Spurr, 1990; Edholm, 1966; Figures 5.1, 5.3, and 5.4).

Another approach is to provide the subject with a portable, battery-operated device for recording activities (Blair & Buskirk, 1987). In this system all data were recorded on a two-channel electrocardiocorder (Delmar Avionics, Irvine, CA), a 6-in. by 4-in. by 1.5-in. reel-to-reel ECG recorder weighing 26 oz with a recording capacity of 26 hr per tape. Rechargeable cadmium batteries powered the unit up to 36 hr. A wristwatch that alerts the subject at specified times to record activities was employed by Riumallo et al., (1989). An investigator can also set the alarm on a face-covered watch and direct the subject to record the activity in which he or she is engaged when the alarm sounds.

Advantages and Disadvantages

There are several advantages to the diary method of estimating physical activity or energy expenditure. Data collection involves little expense and does not require an observer, and data can be collected by many subjects simultaneously. In addition, a list of the actual activities is available for study and provides data to estimate energy expenditure.

On the other hand, there are disadvantages to the diary technique. Processing a large volume of data can be time-consuming and expensive. If energy cost of specific activities is to be determined by measuring oxygen consumption, considerable additional expense and time are involved. Complete cooperation and conscientiousness on the part of the subjects is critical if accurate data are to be obtained. Even so, the subject may forget to log all entries or might make recording errors. Because logging activities is tedious, the longer the period of data collection, the less accurate the results may be. This means that if habitual exercise is to be assessed, shorter periods of recording during various times of the year are necessary. Most persons who have worked with this method agree that it is inappropriate with children below the age of 10 and perhaps even in older children.

If the data from the diary are to be converted to energy expenditure, the energy cost of each activity must be estimated and multiplied by the time period in which the subject was engaged in the activity. This may be as great

INSTRUCTIONS

Fill in your name and other details on Part 1 overleaf.

Fill in details of your activities for all your waking hours using the abbreviations given. This daily diary form covers 24 hours and has a square for each minute.

Try to estimate activity to the nearest minute. For activities not listed use your own abbreviations and give details under 'notes'.

Carry this form with a pencil or pen all day. If in doubt about anything ask your observer.

	0	1	2	3	4	5	6	7	8	9
06.00										
07.00										

ABBREVIATIONS

A	—Ablutions	K	—Kneeling
BM	—Bedmaking	L	—Lying
C	—Cooking	R	—Running
DS	—Dressing	S	—Sitting
FC	—Fieldcraft	SL	—Sleeping
FS	—Floor sweeping	ST	—Standing
I	—Ironing	W	—Walking

Other activities - use own code

NOTES:

Figure 5.1 A diary form for recording one's own activities minute by minute. From Edholm, O.G. (1966). The assessment of habitual activity. In K. Evang & K.L. Andersen (Eds.), *Physical Activity in Health and Disease* (pp. 187-197). Reproduced with permission of the Williams and Wilkins Company, Baltimore, Maryland.

or greater a source of error as the maintenance of the diary. For example, it is well known that less fit subjects will rate activities at higher intensities than will more fit subjects. There are two ways of making these estimates; by using published tables of energy cost (Appendix C) or by measuring oxygen consumption during activities that represent activities recorded in the diary. This second method (often called the factorial method) is thought by some investigators to be more accurate, but the time and expense required make it unsuitable for epidemiologic studies. Also, there is evidence that the measurement of oxygen uptake under these circumstances underestimates energy expenditure of activities recorded by the diary method (Brun, Webb, DeBenoist, & Blackwell, 1985).

If energy costs in kilojoules (kilocalories) reported in the literature are used, they may be inappropriate for young children or older adults if based on an adult population; hence it would be better to use energy cost estimates in terms of METs rather than kilojoules or kilocalories. A combination of these methods has also been used (Blackburn & Calloway, 1976) in which resting metabolic rate is determined in each subject and energy cost tables in METs are used to estimate energy expenditure of activities recorded in the diary. This method is intermediate in cost. A final problem with the diary method is that because subjects know data are being collected, the process itself may result in an alteration of activities.

Validity in Adults

The best studies of the validity of diary records as a measure of habitual physical activity are summarized in Table 5.1. The diary method is quite good for estimating the mean energy expenditure of groups, with an error of not more than 6% and generally about 3% but with a tendency to underestimate energy expenditure slightly. The individual errors are moderately large, about 7% to 8% on the average, but are acceptable for many studies. There appears to be no significant advantage to measuring energy costs for specific activities in each individual compared to using tabled values

Categorical value	Examples of activities	Energy cost in mets from various studies		Median energy cost used	
		Minimum	Maximum	MET	kcal•kg[1] • 15 min^{-1}
1	Sleeping Resting in bed	1.0		1.0	0.26
2	Sitting: eating, listening, writing, etc.	1.0	2.0	1.5	0.38
3	Light activity standing: washing, shaving, combing, cooking, etc.	2.0	3.0	2.3	0.57
4	Slow walk (< 4 km/h), driving, to dress, to shower, etc.	2.0	4.0	2.8	0.69
5	Light manual work: floor sweeping, window washing, driving a truck, painting, waiting on tables, nursing chores, several house chores, electrician, barman, walking at 4 to 6 km/h	2.3	5.0	3.3	0.84
6	Leisure activities and sports in a recreational environment: baseball, golf, volleyball, canoeing or rowing, archery, bowling, cycling (<10 km/h), table tennis, etc.	3.0	8.0	4.8	1.2
7	Manual work at moderate pace: mining, carpentry, house building, lumbering and wood cutting, snow shoveling, loading and unloading goods, etc.	4.0	8.0	5.6	1.4
8	Leisure and sport activities of higher intensity (not competitive): canoeing (5 to 8 km/h), bicycling (>15 km/h), dancing, skiing, badminton, gymnastic, swimming, tennis, horse riding, walking (>6 km/h), etc.	5.0	11	6.0	1.5
9	Intense manual work, high intensity sport activities or sport competition: tree cutting, carrying heavy leads, jogging and running (>9 km/h), racquetball, badminton, swimming, tennis, cross country skiing (>8 km/h), hiking and mountain climbing, etc.	6.0	~ 15	7.8	2.0

(continued)

Figure 5.2 Activity categories and a diary form for recording activities in 15-min intervals. The numbers on the diary form represent a typical day, with summary information also provided. From Bouchard, C., Tremblay, A., LeBlanc, C., Lortie, G., Savard, R., & Thériault, G. (1983). A method to assess energy expenditure in children and adults. Copyright © *American Journal of Clinical Nutrition*, **37**, 461-467. With permission of the American Society for Clinical Nutrition.

CODE OF DAY ____________________

NAME ____________________

SURNAME ____________________

PARENT ☐ CHILD ☐

SEX ____________________

AGE IN YEARS ____________________

DATE ____________________

SUBJECT ID ____________________
Number

Write in the space provided the categorical value which corresponds best to the dominant activity of each 15-minute period. Please, consult the activity card to establish the proper coding. In case of doubt, make a note and raise the problem during the interview.

Hour \ Min.	0-15	16-30	31-45	46-60
0	*2*	*2*	*2*	*1*
1	*1*	*1*	*1*	*1*
2	*1*	*1*	*1*	*1*
3	*1*	*1*	*1*	*1*
4	*1*	*1*	*1*	*1*
5	*1*	*1*	*1*	*1*
6	*1*	*1*	*1*	*1*
7	*1*	*1*	*1*	*1*
8	*1*	*1*	*1*	*1*
9	*3*	*4*	*3*	*4*
10	*5*	*5*	*5*	*5*
11	*5*	*5*	*5*	*5*
12	*5*	*5*	*5*	*2*
13	*2*	*2*	*5*	*5*
14	*5*	*5*	*5*	*5*
15	*5*	*5*	*2*	*2*
16	*2*	*2*	*2*	*2*
17	*2*	*2*	*4*	*2*
18	*2*	*2*	*4*	*4*
19	*2*	*2*	*2*	*2*
20	*2*	*2*	*2*	*2*
21	*8*	*8*	*8*	*4*
22	*4*	*4*	*2*	*2*
23	*2*	*2*	*2*	*1*

Summary:			
	1 = *34*	4 = *8*	7 = *0*
	2 = *30*	5 = *19*	8 = *3*
	3 =*2*	6 =*0*	9 = *0*

Figure 5.2 *(continued)*

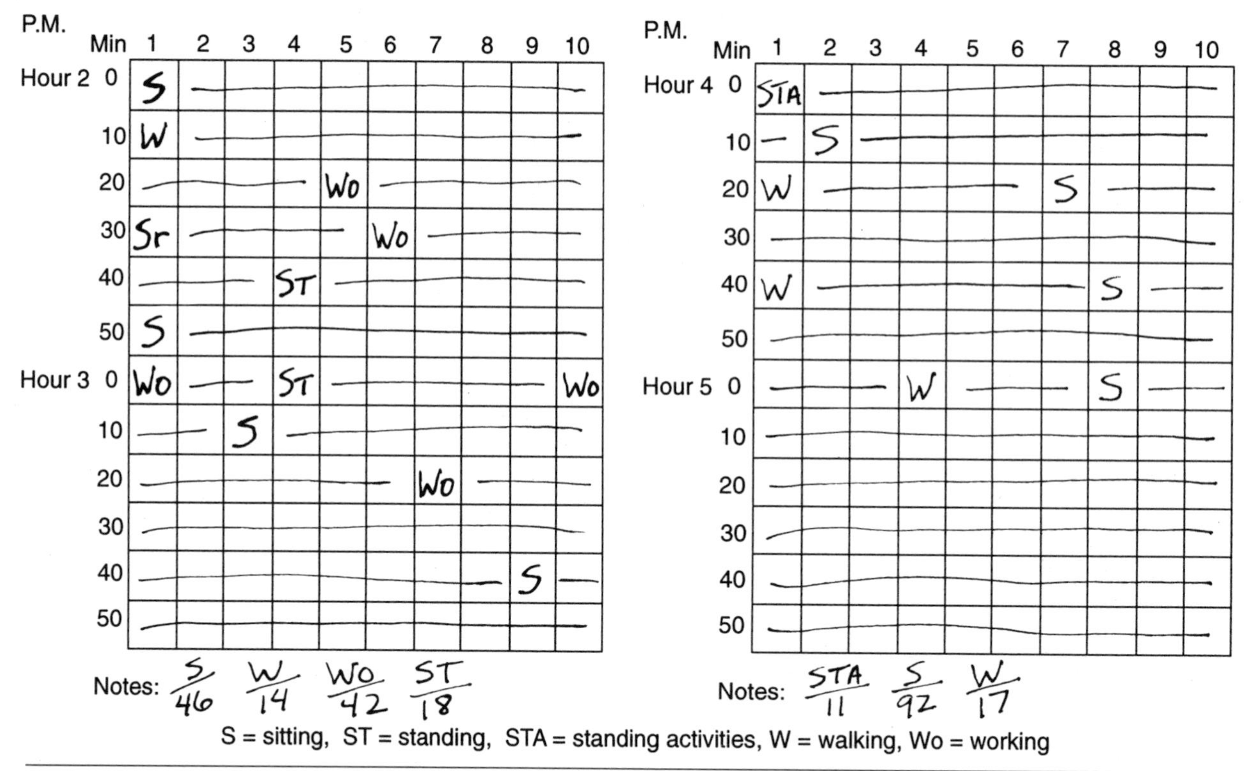

Figure 5.3 A diary form for recording changes in physical activities. From Durnin, J.V.G.A., & Passmore, R. (1967). *Energy, Work and Leisure*, p. 28, with permission of J.V.G.A. Durnin.

of energy costs for those activities. It should be recognized that these studies all involve a small number of subjects under somewhat artificial conditions, dictated by the sophisticated criteria of energy expenditure that are employed. Hence, the application of the results to larger populations of free-living people is open to question. And although these criteria are as good as any available at this time, there is still the possibility of some error in the criterion value of energy expenditure.

Results of some other validation studies (Bouchard et al., 1983; Leclerc, Allard, Talbot, Gauvin, & Bouchard, 1985; Curtis & Bradfield, 1971; Marable, Kehrberg, Judd, Prather, & Bodwell, 1988) are not listed in Table 5.1 because the criteria used for their evaluation are questionable.

Validity in Children

Unfortunately, validity data for children comparable to the data of Table 5.1 for adults are limited. In one investigation (Bouchard et al., 1983), data from 150 children were pooled with those of 150 adults, the entire age range being 10 to 50 years. The diary method required the subjects to record their activities every 15 min for 3 days. Literature values of energy cost of activities were used in calculating daily energy expenditure. The data collection was repeated during nine periods of 3 days each throughout the year. The diary technique was reliable in that the 3-day averages, when compared among the nine periods, resulted in an intraclass correlation coefficient of .88. Unfortunately, body weight was not partialed out in any way. As mentioned earlier, total energy expenditure when expressed in kilojoules or kilocalories is influenced considerably by body weight, which is a confounding variable if one is to interpret kilojoules or kilocalories in terms of physical activity.

SUMMARY

The diary method requires complete cooperation and good judgment from the subject, so the technique is not practical with some populations, such as young children. In some instances, it is not reasonable to expect subjects to frequently interrupt their activities to record data. Data collection is inexpensive, because many subjects can be keeping diaries simultaneously and an observer is not required. The method is subject to considerable individual error (Table 5.1), although the accuracy is sufficient for group estimates. The errors result primarily from errors in recording (a tedious procedure), inaccurate estimates of the strenuousness of activities, and errors in the tabular energy cost values for various activities.

Table 5.1 Validation of the Diary Method: Summary of Studies

Subjects	Recording interval	No. of days of recording	O_2 uptake of activities measured	Criterion	Accuracy and precision (%): Mean criterion minus mean diary method	Mean of absolute individual differences	Range of individual differences	*r* or rho	Reference
6 males (prefeeding)	1 min	1 week	No	Doubly labeled water	5.6	8.8	−4 to 20	.7*	Riumallo et al., 1989
			Yes		5.5	9.7	−12 to 15	1.0*	
5 males (postfeeding)	1 min	1 week	No		1.9	2.6	−1 to 6	.9*	
			Yes		1.0	2.7	−3 to 6	.8*	
6 males	3 min	5 days	Measured basal metabolic rate	Dietary intake food weighed	−1.6	7.2	−11 to 15	.90*	Borel, Riley, & Snook, 1984
6 females	3 min	5 days			4.7	8.4	−11 to 16	.53	
11 males	1 min	24 hr	Yes	Room calorimeter, VO_2	−3	±9 (SD)	−12 to 20	—	Geissler, Dzumbira, & Noor, 1986
14 females	1 min	24 hr	Yes		1	±9 (SD)	−17 to 25	—	
Total subjects	1 min	24 hr	Yes		−1	±9 (SD)	−17 to 25	—	
12 males	1 min	2 weeks	Yes	Dietary intake food weighed	−0.5	9.7	−15 to 17	—	Edholm & Fletcher, 1955
6 males	15 min	2 weeks	Resting metabolic rate	Doubly labeled water	−4.2	12.4	−20 to 21	0.57	Schulz, Westerterp, & Brück, 1989
			Basal metaoblic rate		−7.3	11.3	−25 to 23	0.72*	
12 females	10 min	5 days	No	Dietary intake food weighed	−6		−39 to 56	—	Kalkwarf, Haas, Belko, Roach, & Roe, 1989
			Yes		−2		−34 to 37	—	
5 males	1 min	13 days	Yes	Dietary intake food weighed	3.1		−7 to 10	—	Passmore, Thompson, & Warnock, 1952
12 males	5 min	35-315 days	No	Dietary intake	0.3 (−6.7)[a]	15.0 (9.4)[a]	−18 to 77	—	Acheson, Campell, Edholm, Miller, & Stock, 1980
			Yes	Bomb calorimeter	5.9 (−0.2)[a]	15.0 (9.8)[a]	−17 to 73	—	

[a]Value in parenthesis is the value with one subject excluded. His diary value was more than 70% above criterion value.
*Statistically significant, $p < 0.05$.

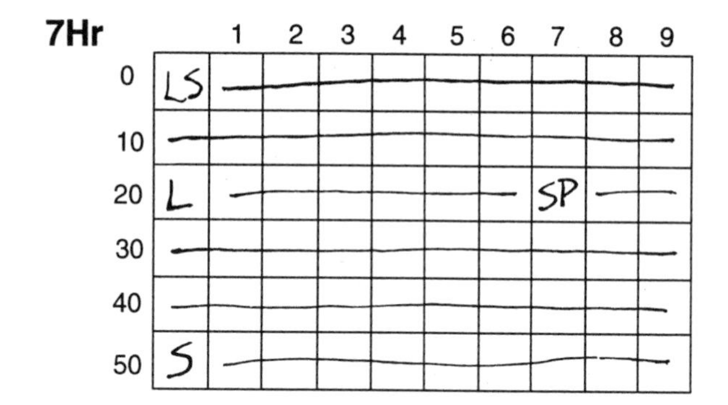

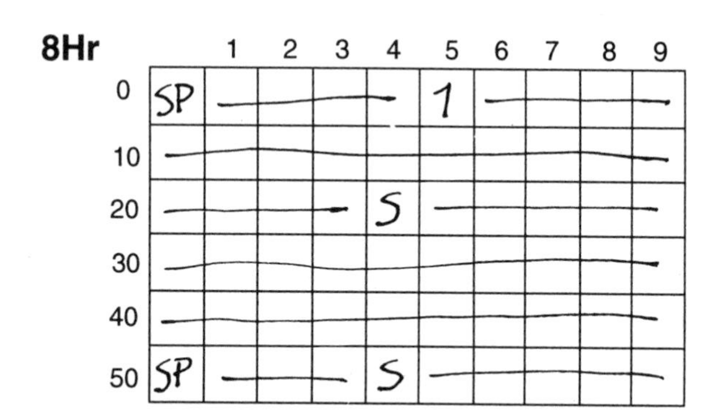

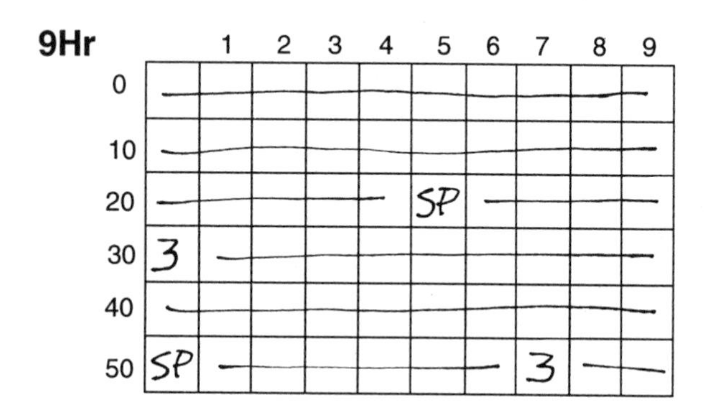

Figure 5.4 A schoolboy's sample diary of activity between 7:00 a.m. and 10:00 a.m., using this code: LS, lying down, asleep; L, lying down, awake (reading, talking); S, sitting or standing still; SP, standing, moderate activity, walking slowly; 1, light muscular activity (e.g., fast walking, climbing stairs); 2, moderate muscular activity (e.g., running, cycling, gymnastics); 3, heavy muscular activity (e.g., football, basketball, physical training): X, other activities that are difficult to describe. From Andersen, K.L., Masironi, R., Rutenfranz, J., & Seliger, F. (1978). *Habitual Physical Activity and Health*, Copenhagen: World Health Organization (European Series No. 6). Reproduced with permission.

REFERENCES

Acheson, K.J., Campell, I.T., Edholm, O.G., Miller, D.S., & Stock, M.J. (1980). The measurement of daily energy expenditure: An evaluation of some techniques. *American Journal of Clinical Nutrition*, **33**, 1155-1164.

Andersen, K.L., Masironi, R., Rutenfranz, J., & Seliger, F. (1978). *Habitual physical activity and health.* Copenhagen: World Health Organization Regional Publication. (European Series No. 6)

Blackburn, N.W., & Calloway, D.H. (1976). Energy expenditure and consumption of mature, pregnant and lactating women. *Journal of the American Dietetic Association*, **69**, 29-37.

Blair, D., & Buskirk, E.R. (1987). Habitual daily energy expenditure and activity levels of lean and adult-onset and child-onset obese women. *American Journal of Clinical Nutrition*, **45**, 540-550.

Borel, M.J., Riley, R.E., & Snook, J.T. (1984). Estimation of energy expenditure and maintenance energy requirements of college-age men and women. *American Journal of Clinical Nutrition*, **40**, 1264-1272.

Bouchard, C., Tremblay, A., Leblanc, C., Lortie, G., Savard, R., & Thériault, G. (1983). A method to assess energy expenditure in children and adults. *American Journal of Clinical Nutrition*, **37**, 461-467.

Brun, T., Webb, P. DeBenoist, B., & Blackwell, F. (1985). Calorimetric evaluation of the diary-respirometer technique for the field measurement of the 24-hour energy expenditure. *Human Nutrition and Clinical Nutrition*, **39C**, 321-334.

Collins, K.J., & Spurr, G.B. (1990). Energy expenditure and habitual activity. In K.J. Collins (Ed.) *Handbook of methods for the measurement of work performance, physical fitness and energy expenditure in tropical populations* (pp. 81-90). Paris: International Union of Biological Sciences.

Curtis, D.E., & Bradfield, R.B. (1971). Long-term energy intake and expenditure of obese housewives. *American Journal of Clinical Nutrition*, **24**, 1410-1417.

Durnin, J.V.G.A. (1967). Activity patterns in the community. *Canadian Medical Association Journal*, **96**, 882-886.

Durnin, J.V.G.A., & Passmore, R. (1967). *Energy, work and leisure.* London: Heinemann Educational Books.

Edholm, O.G. (1966). The assessment of habitual activity. In K. Evang & Andersen, K.L. (Eds.), *Physical activity in health and disease* (pp. 187-197). Oslo: Williams and Wilkins. (Proceedings of the Betosölen Symposium)

Edholm, O.G., & Fletcher, J.G. (1955). The energy expenditure and food intake of individual men. *British Journal of Nutrition*, **9**, 286-300.

Geissler, C.A., Dzumbira, T.M.O., & Noor, M.I. (1986). Validation of a field technique for the measurement of energy expenditure: Factorial method versus continuous respirometry. *American Journal of Clinical Nutrition,* **44,** 596-602.

Gorsky, R.D., & Calloway, D.H. (1983). Activity pattern changes with decreases in food energy intake. *Human Biology,* **55,** 577-586.

Kalkwarf, H.J., Haas, J.D., Belko, A.Z., Roach, R.C., & Roe, D.A. (1989). Accuracy of heart-rate monitoring and activity diaries for estimating energy expenditure. *American Journal of Clinical Nutrition,* **49,** 37-43.

LaPorte, R.E. (1979). An objective measure of physical activity for epidemiologic research. *American Journal of Epidemiology,* **109,** 158-168.

Leclerc, S., Allard, C., Talbot, J., Gauvin, R., & Bouchard, C. (1985). High density lipoprotein cholesterol, habitual physical activity and physical fitness. *Atherosclerosis,* **57,** 43-51.

Marable, N.L., Kehrberg, N.L, Judd, J.T., Prather, E.S., & Bodwell, C.D. (1988). Caloric and selected nutrient intakes and estimated energy expenditures for adult women: Identification of non-sedentary women with lower energy intakes. *Journal of he American Dietetic Association,* **88,** 687-693.

Riumallo, J.A., Schoeller, D., Barrera, G., Gattas, V., & Vauy, R. (1989). Energy expenditure in underweight free-living adults: Impact of energy supplementation as determined by doubly labeled water and indirect calorimetry. *American Journal of Clinical Nutrition,* **49,** 239-246.

Schulz, S., Westerterp, K.R., & Brück, K. (1989). Comparison of energy expenditure by the doubly labeled water technique with energy intake, heart rate, and activity. *American Journal of Clinical Nutrition,* **49,** 1146-1154.

Chapter 6

QUESTIONNAIRES AND INTERVIEWS

Questionnaires and interviews seeking information about habitual physical activity may take several forms. A prepared questionnaire can be self-administered—for example, by being sent or delivered to a respondent, who answers the questions without assistance. Or an interviewer using a prepared form can record answers to questions about activity, either in person or by telephone. A combination of self-administered questionnaire and interview has also been used; the respondent completes a questionnaire and, after studying the responses, the interviewer probes to either clarify or cross-check responses.

A naive investigator or practitioner can readily be confused by the proliferation of questionnaires that have been used to assess physical activity. Only in rare instances has the validity or reliability of the procedures been studied. Some researchers simply describe the questionnaires they have designed and used. We will confine our attention to the most popular questionnaires and interviews that appear to be most useful and valid, and we will only occasionally refer to other forms and procedures.

In some surveys, only occupational activity has been of interest, in others, only leisure time exercise; but many have sought information about activity both on and off the job. Questionnaires and interviews have been used to assess physical activity for various periods—the previous 24 hr, the previous week, the previous 3 months, the previous year. Some attempt to obtain activity profiles during respondents' entire lifetimes.

Organizing an Approach

The specific purpose of the survey will determine the procedures and questions to be asked. For example, if obesity were of interest, total energy expenditure might be estimated. For an investigation concerned with osteoporosis and hip fractures, questions about weight-bearing activities might be sufficient. Or if maximal cardiovascular fitness (i.e., aerobic power) was the concern, then it might be important to determine participation in high-intensity, large-muscle aerobic exercise.

The procedures and form of the questionnaire or interview might be different if one wished to classify respondents into three or four activity groups, versus wishing to obtain a specific activity score or energy expenditure for each individual. A few simple survey questions require little time or expense, but they may not provide the information needed. The more detail required, the more time and expense is involved.

Questionnaires should request objective rather than subjective responses when possible to provide consistent interpretation and scoring. Demographic data (age, weight, sex, etc.) should also be obtained for use in analysis.

Advantages and Limitations

There are a number of advantages to the questionnaire/interview technique compared to other approaches (observation, heart rate recording, diaries, etc.). It is relatively inexpensive, and at present it is the only method feasible for large population surveys. Specific activities can be identified together with frequency and duration. The procedure doesn't influence subjects' activities to the extent that can occur with observation or diary keeping. By employing energy expenditure tables, it may be possible to estimate total energy expenditure. The procedure is usable over a wide age range, and many subjects enjoy describing their activities.

But there are limitations to the method as well. Subjects do not necessarily recall their activities accurately; they may tend to overestimate time or intensity. A self-administered questionnaire must be suited to respondents' ages and education levels. Detailed questionnaires and interviews place a considerable burden on subjects. For example, in one of our studies we had to abandon questionnaire and interview assessments with women because the time required to obtain details of household chores annoyed respondents, and we feared other aspects of the study might be jeopardized. But there is evidence that activity is underestimated in women when household chores are not included (Ainsworth & Leon, 1991; Shaw, 1985).

If the activity data are to be converted to energy expenditure, it is necessary to use published tables, which are based mostly on male adults and may not reflect the energy costs of activities for other populations. There is a wide range in energy cost for many activities depending on how they are performed, and many subjects have difficulty estimating the intensity (strenuousness) with which they engage in activities. For this reason some questionnaires and interviews ask for information about sweating or breathlessness. Perhaps a subjective rating of exertion, such as the Borg Scale (Borg, 1982), might improve some of the questionnaire/interviews if incorporated into the survey. There is considerable evidence that strenuous activity is recalled more accurately than moderate or mild activity (Taylor et al., 1984).

As with other methods of assessment, validity with the questionnaire/interview technique is difficult to determine because of inadequate criteria. Doubly labeled water is an acceptable criterion, but this is limited to average energy expenditure over a limited time frame. Careful observation, calorimetry, and diaries might also serve as validation criteria. Energy intake with food analyzed or at least weighed has been shown to be a good criterion for energy expenditure. All of these criteria are appropriate only for small numbers of respondents.

Sometimes the results of other methods (dietary recall, heart rate recording, movement counters, etc.) are used as criteria, but the error rates may be as great as in the questionnaire or interview. Validation by these methods is included in our discussion, but the reader should attach less credence to these data. Some studies have attempted to validate questionnaires or interviews against physical fitness (aerobic power, for example) or body fatness. Although it is true people who are more active tend to be more fit and more lean, these characteristics are influenced in large measure by other factors, such as heredity. Hence, these validation studies are also of limited value. Finally, if an activity questionnaire or interview can predict, even to a limited extent, who is healthy and who is not, the method can be said to have face validity—it works. Nevertheless, this criterion, too, is superficial. If no relationship is found between, say, a disease risk factor and activity, is this because there truly is no relationship or because the method of assessing physical activity is inaccurate?

This general discussion of questionnaires and interviews applies to subjects of all ages. However, there are additional problems and considerations in using this approach with the elderly and with children, and we have included separate sections in this chapter for young and middle-aged adults who are generally employed, elderly adults, and children.

Young and Middle-Aged Adults

Some investigators wishing to study the relationship of physical activity to health or disease have compared two occupational cohorts that differed widely in occupational physical activity. This is a crude way of classifying subjects on the basis of habitual physical activity. Examples are the early classic studies of London bus drivers and conductors by Morris, Heady, Raffle, Roberts, and Parks (1953), of longshoremen by Paffenbarger, Laughlin, Gima, and Black (1970), and of railroad workers by Taylor et al. (1962). Classifications such as these generally don't involve questionnaires or interviews, but these investigations have nonetheless produced valuable information; however, discussion of the results are outside the scope of this book.

Job Classification

The assessment of occupational physical activity or energy expenditure among people of a variety of occupations is a different matter. Questionnaires and interviews are convenient and inexpensive tools for classifying employed men and women by physical activity. They are acceptable to the subjects and do not generally influence work tasks. One must allow for the fact that some people work two or three jobs. Several other disadvantages of job classification are that

- some occupations involve a multitude of tasks of varying strenuousness;

- it does not take into account leisure time physical activity, including going to and from work;
- advanced technology brings continual changes in occupational physical activity;
- job titles can be misleading in estimating physical activity of work;
- there are seasonal changes in many jobs; and
- the method is unusable with homemakers, children, the unemployed, or retirees.

The distinguished anthropologist Raymond Pearl saw the importance of assessing physical activity and developed one of the first methods of classifying workers on the basis of the strenuousness of their occupations (Pearl, 1924). In an exhaustive study, he had nine raters classify occupations numerically from 1 to 5 according to the energy demands. The mean occupational scores were in turn divided into quintiles to arrive at five classes ranging from sedentary to very heavy. Other attempts to classify occupations on the basis of energy expenditure followed (Brouha, 1960; Christensen, 1953; Lehman, 1953; Purdue Farm Cardiac Project, 1961). In 1955, the United States Employment Service rated and classified 4,000 occupations according to physical demands (U.S. Employment Service, 1955). This scale was modified and condensed 2 years later (National Office of Vital Statistics, 1957). Both reports were related to the numerical classifications in the *Dictionary of Occupational Titles* (Division of Occupational Analysis, 1949). Unfortunately, these publications are of limited use in estimating habitual physical activity today, because job titles are misleading and occupations have changed greatly since the reports were written.

Powell, Thompson, Caspersen, and Kendrick (1987) pointed out in their review that in 20 of the 24 studies of the relationship of physical activity to coronary heart disease conducted before 1970, only occupational activity was examined. Conversely, only 3 of 19 investigations since 1970 examined occupational activity alone. This reversal is due partly to the decline in physical activity on the job resulting from advanced technology. It has become clear that the only source of physical activity for many people is leisure time, including the time spent going to and from work.

Many earlier studies based solely on occupational physical activity established a link between physical activity and disease, particularly coronary heart disease. One of the first to do so was the report by Hedley (1939). Powell et al. (1987) review many of these early studies.

Tecumseh and Minnesota Questionnaire/Interviews

For an epidemiologic investigation of the inhabitants of an entire community (Tecumseh, Michigan), a means of assessing habitual physical activity during the previous year was needed (Montoye, 1975). A self-administered questionnaire, based on an earlier experience (Wessel, Montoye, & Mitchell, 1965) was developed and tested in a 10% sample of adults (about 350 people) in the community. There was considerable evidence of misunderstanding among many subjects, and many questions were not answered or answered incompletely. The investigator concluded that a personal interview was necessary in a total population of people with varying occupations and educational achievement (Montoye, 1971). It was also decided not to attempt to assess women's habitual physical activity because of the complexity of household activities for many.

The questionnaire was revised and interview forms were developed. This version was field-tested in another 10% of the adult males, after which the procedure was again revised and an objective scoring system devised. The final procedure and forms that estimate physical activity during the past year were published (Reiff et al., 1967). Although many occupations in industrialized countries are sedentary, the Tecumseh study showed that for the males in an entire community, occupational activity contributes more to total activity than does leisure activity (Montoye, 1971). Of course, among people with sedentary jobs, leisure time is the only source of significant physical activity. The Tecumseh questionnaire/interview procedure has been used to study the relationship of coronary risk factors and habitual physical activity and also has been used in other analyses (Cunningham, Montoye, Metzner, & Keller, 1968, 1969; Montoye, Block, Metzner, & Keller, 1976, 1977; Montoye, Metzner, Keller, Johnson, & Epstein, 1972; Montoye, Mikkelsen, Metzner, & Keller, 1976).

The occupational part of the Tecumseh questionnaire/interview was recently revised by David Jacobs and Henry Montoye (1988). This revision, including directions for the interview, appears in Appendix D. The strenuous ratings for all activities are in METs, a ratio of exercise metabolic rate divided by resting metabolic rate. MET values of other activities can be found in Appendix C. The leisure time activity part of the Tecumseh procedure was revised and improved by Taylor and colleagues (1978) and more recently by David Jacobs (1988). The most recent revision (called the Minnesota Leisure Time Activity Questionnaire) and directions are also contained in Appendix D. The MET values to be used with this questionnaire/interview are shown. Again, values from Appendix C can be used if necessary. The total activity score, combining occupation, leisure time, and transportation to and from work, is expressed as an average MET value per week. A sample scoring sheet showing the necessary calculations also appears in Appendix D. Of course, a computer program can be developed to do the calculations. The total administration time required for the occupational plus the leisure time questionnaire/interview averages about 1 hr.

Validity. The original, combined occupational and leisure time questionnaire/interview was carefully tested in a total community. Comparing energy intake determined by 24-hr dietary recall, the most active 20% of the 883 males consumed on the average about 13% more energy than the least active 20%, and the intermediately active group consumed about 10% more energy than the least active group (Montoye et al., 1972). Assuming caloric balance, this represents a crude validation of the questionnaire/interview. In several investigations the Tecumseh or Minnesota assessment (or both) has been compared to physical activity as measured by other questionnaires/interviews; the results are shown in Table 6.1. A comparison with devices for monitoring physical activity is shown in Table 6.2 and with various measures of physical fitness in Table 6.3. None of these other measures of physical activity should be considered infallible criteria because their errors may be as great or greater than the Tecumseh and Minnesota questionnaire/interviews. These results are presented for want of a definitive standard. Leon, Connett, Jacobs, and Rauramaa (1987), in a 7-year follow-up of 12,138 men, demonstrated that scores on the Minnesota Leisure Time Activity Questionnaire/Interview were related to subsequent development of coronary heart disease. The most active third of the men at initial assessment had a third fewer coronary events during the subsequent 7 years than the least active third of the population. The experience of the middle third was comparable to the most active third. A 10-1/2-year follow-up of the group produced similar results (Leon, Connett, and the MRFIT Research Group, 1991). In both analyses adjustments had been made for other risk factors. If one is studying coronary heart disease, these results serve as one measure of validity.

Reliability. The reliability of the Minnesota Leisure Time Activity Questionnaire has received some attention. Folsom, Jacobs, Caspersen, Gomez-Marin, and Knudsen (1986) reported that the results of one administration correlated with the results of a second administration 1 month later ($r = .88$). Blair et al. (1991) administered the Minnesota questionnaire to 129 males and 322 females. Some of these subjects were then asked 1 year to 4 years later to report what their leisure activities were at the previous administration. There was moderate agreement in the two assessments (r varied from .33 to .52). As we might expect, recall was poorer as the intervening period lengthened. Jacobs, Ainsworth, Hartman, and Leon (1993), in a study with 77 adults, reported a test-retest coefficient for the Tecumseh occupational questionnaire of .92 with 1 week between administrations and .69 with 12 weeks intervening.

Paffenbarger/Harvard Alumni Questionnaire

The Tecumseh and Minnesota questionnaire/interview procedures solicit details about physical activity and require considerable subject and interviewer time, but they are suitable for respondents of a wide educational range. Ralph Paffenbarger and colleagues devised a much simpler self-administered questionnaire to study the relationship of physical activity and disease among a population of Harvard University alumni. The questions ask about stairs climbed, city blocks walked, sports activities, light activities, and the like. A recent modification from Paffenbarger, Blair, Lee, and Hyde (1993) is reproduced as Appendix E., together with the method of calculating kilocalories per week in leisure time activities (Lee, 1992).

Validity. The Paffenbarger questionnaire has been used effectively in illustrating that leisure time physical activity is inversely related to the incidence of several diseases and thus can be said to have face validity (Lee, Hsieh, & Paffenbarger, 1993; Paffenbarger, 1985; Paffenbarger, Hyde, & Wing, 1987; Paffenbarger, Hyde, Wing, & Hsieh, 1986; Paffenbarger, Hyde, Wing, Jung, & Kampert, 1991; Paffenbarger, Hyde, Wing, & Steinmetz, 1984; Paffenbarger, Wing, & Hyde, 1978; Paffenbarger, Wing, Hyde, & Jung, 1983). In a large general population, Washburn, Smith, Goldfield, and McKinlay (1991) reported low but statistically significant correlation coefficients between the Harvard alumni questionnaire results and high-density lipoprotein cholesterol (positive) and body mass index (negative). The relationships between the Paffenbarger/Harvard alumni index of physical activity and other questionnaires, activity monitoring devices, and $\dot{V}O_2max$ are shown in Tables 6.4 and 6.5.

Reliability. The reliability of the Paffenbarger questionnaire has received only limited attention. In 14 subjects the test-retest correlation coefficient was .76 after a 4-week interval (Cauley, LaPorte, Black-Sandler, Schramm, & Kriska, 1987). After a year these authors reported a coefficient of .73 in 59 subjects. Jacobs, Ainsworth, Hartman, and Leon (1993) reported a coefficient of .50 with 78 adults with 8 weeks separating the two administrations of the questionnaire. In a large sample of men and women, test-retest correlation with 7 to 12 weeks intervening resulted in coefficients of .58 for the entire population and .69 for those whose activity patterns did not change (Washburn, Smith, Goldfield, & McKinlay, 1991).

Five-City/7-Day Recall Questionnaire/Interview

The 7-day physical activity recall questionnaire was originally developed in 1979 for use in the Stanford

Table 6.1 Physical Activity as Measured by the Tecumseh and/or Minnesota Leisure Time Questionnaire/Interviews Compared to Other Questionnaires/Interviews

Tecumseh and/or Minnesota questionnaires	Other questionnaire method[a]	Population	Correlation coefficient	Reference
Tecumseh occupational and leisure time METs	HIP[b] occupational and leisure time	197 males	.49*	Buskirk, Harris, Mendez, & Skinner, 1971
Tecumseh occupational METs	HIP occupational	197 males	.52*	
Tecumseh occupational METs	Saltin and Grimby occupational	43 females	.45*	Wilbur et al., 1989
Minnesota leisure time METs	Saltin and Grimby leisure time	43 females	.09	
Minnesota leisure time METs	4-week histories (14 administrations)	73 males and females	.74*	Jacobs et al., 1993
Minnesota leisure time kcal	Paffenbarger	21 males	.54*	Albanes et al., 1990
	7-Day recall		.48*	
	Framingham		.33	
	Baecke		.36*	
	Lipid Research Clinic		.63*	
	HIP		.52*	

[a]Other methods are discussed later in the chapter.
[b]HIP = Health Insurance Plan of New York questionnaire.
*Statistically significant, $p < .05$.

Table 6.2 Physical Activity as Measured by the Tecumseh and/or Minnesota Leisure Time Questionnaire/Interviews Compared to Doubly Labeled Water, Activity Monitoring, and Caloric Intake

Tecumseh and/or Minnesota questionnaires	Other assessment method	Population	Correlation coefficient	Reference
Tecumseh occupational METs	Beeper occupational METs[a]	43 females	.29*	Wilbur et al., 1989
Minnesota leisure time METs	Beeper leisure time METs[a]	43 females	.14	
Minnesota leisure time METs	LSI monitor[b]	42 males	.45*	LaPorte et al., 1982
Minnesota leisure time kcal	Measured kcal intake less resting	21 males	.17	Albanes et al., 1990
Minnesota leisure time METs (14 administrations)	Caltrac accelerometer[b] (METs; 14 two-day administrations)	73 males and females	.18	Jacobs et al., 1993
	Caltrac accelerometer (kcal)		−.06	
Tecumseh occupational and Minnesota leisure time METs	Doubly labeled water (METs)	28 males	.26	Gretebeck, Montoye, & Porter, 1993
	Caltrac accelerometer[b] (METs; 7 days)		.40*	
	Energy intake (METs; 7 days)		−.06	

[a]When beeper sounded, subject recorded activity.
[b]To be discussed in chapter 7. LSI = Large-Scale Integrated Motor Activity Monitor.
*Statistically significant, $p < .05$.

Table 6.3 Physical Activity as Measured by the Tecumseh and/or Minnesota Leisure Time Questionnaire/Interviews Compared to Measures of Physical Fitness

Tecumseh and/or Minnesota questionnaires	Measure of physical fitness	Population	Correlation coefficient	Reference
Minnesota leisure time METs	Treadmill endurance	175 males	.45*	Taylor et al., 1978
Tecumseh occupational and leisure time	Submaximal treadmill test	54 males	.13 .21	Skinner, Benson, McDonough, & Hames, 1966
Tecumseh occupational Minnesota leisure time kcal	Treadmill endurance submaximal treadmill test (heart rate)	175 males	.41* .59*	Leon, Jacobs, DeBacker, & Taylor, 1981
Minnesota leisure time METs	Submaximal treadmill test	1,513 males	.10*	DeBacker et al., 1981
Minnesota leisure time METs (14 administrations)	$\dot{V}O_2max$ submaximal treadmill test	64 males and females	.43* .45*	Jacobs et al., 1993

*Statistically significant, $p < .05$.

Table 6.4 Paffenbarger/Harvard Alumni Questionnaire Index Compared to Physical Activity as Measured by Other Questionnaires/Interviews

Other questionnaires	Population	Correlation coefficient	Reference
Six 48-hr activity records	78 adults	.33*	Ainsworth, Jacobs, & Leon, 1990
7-Day recall,[a] kcal	21 males	.09	Albanes et al., 1990
Framingham,[a] kcal		.36*	
Minnesota leisure time kcal		.54*	
Baecke index*[a]		.56*	
Health Insurance Plan of New York index[a]		.53*	
Lipid Research Clinics[a]		.81*	
Baecke leisure time index[a]	130 postmenopausal females 125 postmenopausal females	.20* .16*	Cauley et al., 1987
Paffenbarger sweat frequency[b]	36 males 32 females	.57*	Siconolfi, Lasater, Snow, & Carleton, 1985
Paffenbarger sweat frequency[b]	732 males and females	.39*	Washburn, Goldfield, Smith, & McKinlay, 1990
Minnesota leisure time (METs)	78 males and females	.31*	Jacobs et al., 1993

[a]Discussed later in the chapter.
[b]Response to frequency of sweating per week as a result of exercise (Paffenbarger question).
*Statistically significant, $p < .05$.

Table 6.5 Paffenbarger/Harvard Alumni Questionnaire Results Compared to Measurement by Doubly Labeled Water, Activity Monitoring, Caloric Intake, and $\dot{V}O_2max$

Other assessment method	Population	Correlation coefficient	Reference
Doubly labeled water, METs	28 males	.39*	Gretebeck et al., 1993
Caltrac accelerometer[a], METs (7 days)		.70*	
Energy intake, METs (7 days)		.44*	
Large-Scale Integrated Motor Activity Monitor[a]	76 females, ages 44-77	.23*	LaPorte et al., 1982
Large-Scale Integrated Motor Activity Monitor[a]	255 postmenopausal females	.11*	Cauley et al., 1986
Large-Scale Integrated Motor Activity Monitor[a]	130 postmenopausal females	.33*	Cauley et al., 1987
	125 postmenopausal females	.17*	
Caloric intake minus resting	21 males	.32	Albanes et al., 1990
Caltrac accelerometer,[a] kcal (14 two-day sessions)	73 males and females	–0.03	Jacobs et al., 1993
Caltrac accelerometer,[a] METs		.30*	
$\dot{V}O_2max$	36 males	.29*	Siconolfi et al., 1985
$\dot{V}O_2max$	32 females	.46*	
$\dot{V}O_2max$	64 males and females	.52*	Jacobs et al., 1993
Submaximal (treadmill)		.52*	

[a]To be discussed in chapter 7.
*Statistically significant, $p < .05$.

Five-City Project (Blair et al., 1985; Sallis et al., 1985). A questionnaire was needed that took less that 15 min to administer, included work-related as well as leisure time activities, estimated total energy expenditure, and applied to all adults. This instrument was selected because the Stanford investigators believed that recall of selected activities from the preceding week should be more accurate than general recall. It was felt that a shorter time, such as a 2- or 3-day recall, would provide very unstable estimates of usual activity (this decision was primarily based on earlier studies of dietary recall). On the other hand, it appeared very difficult to accurately recall the details of specific activities beyond a week. An interviewer-administered format was selected to allow probing for specific activities and to verify statements regarding type, intensity, and duration of exercise.

Participants are asked about the number of hours spent in sleep and in moderate, hard, and very hard activities during the previous week. A prompting list provides examples of activities in each category. The week is separated into weekend days and weekdays. The amount of time remaining each day is assumed to have been spent performing light activities. Because most adults in developed countries spend the greatest part of their days in light activity, it takes little time to recall the time spent in moderate, hard, or very hard activity. Thus, a person need remember only those activities that were at least moderate in energy cost. The more vigorous the activity, the more likely it seems that the specific details will be remembered. The form used is illustrated in Appendix F. The method has been modified slightly (Gross, Sallis, Buono, Roby, & Nelson, 1990; Sallis, Patterson, Buono, Atkins, & Nader, 1988), and an interview manual has been developed by Joan Campbell (James Sallis, personal communication, 1993). If an assessment of habitual physical activity is desired, one must assume that the previous week, on which data is collected, is typical of all weeks, or the procedure must be repeated at various times during the year and the results averaged.

Validity. Comparisons with other methods of activity assessment are shown in Table 6.6.

Reliability. Reliability of the 7-day recall questionnaire and interview has received some attention. With a 2-week separation, Sallis et al. (1985) with 53 adults reported a reliability coefficient of .67 for kilocalories per day but only .34 when expressed as kilocalories per day per kilogram of body weight. They noted that vigorous activity is recalled more reliably. In a later study with 45 adults, administered twice, 1 week apart and expressed as kilocalories per day, the reliability coefficients were .77 and .84 (Williams, Klesges, Hanson, & Eck, 1989). For the total score with 78 adults, again 1 week intervening, a reliability coefficient of .34 was reported (Jacobs et al., 1993). In 44 subjects who

Table 6.6 Five-City/7-Day Questionnaire/Interview Results Compared to Physical Activity as Measured by Doubly Labeled Water, Other Questionnaires, Monitoring Devices, Caloric Intake, and Physical Fitness of Respondents

Other assessment method	Population	Mean $kcal \cdot kg^{-1} \cdot day^{-1}$		Correlation coefficient	Reference
		Diary	7-Day recall		
7-day diary	30 males	39.6	37.7	.81*	Taylor et al., 1984
New Zealand questionnaire	140 males and females			.30*	Hopkins et al., 1991
Three 1-hr observations	44 males and females			.47* to .70*	Klesges et al., 1990
Daily log[a]	45 males and females			.68* to .82*	Williams et al., 1989
Doubly labeled water, METs	28 males			.30*	Gretebeck et al., 1993
Caltrac accelerometer,[b] METs (7 days)				.49*	
Energy intake, METs (7 days)				.23	
Caloric intake, kcal	21 males			.35*	Albanes et al., 1990
Paffenbarger/Harvard questionnaire				.09	
Framingham questionnaire[c]				.45*	
Minnesota leisure time questionnaire				.48*	
Baecke questionnaire[c]				.16	
Health Insurance Plan of New York questionnaire[c]				.40*	
Lipid Research Clinics questionnaire[c]				.31	
Caltrac[b] (movement score)	31 males and females			.14 to .37*	Williams et al., 1989
Caltrac accelerometer[b] (movement score)	45 males and females			.57*	Rauh, Hovell, Hofstetter, Sallis, & Gleghorn, 1992
Treadmill endurance	265 males			.17*	Dishman, 1988
Caltrac accelerometer,[b] kcal (7 days)	7 males and 26 females			.79*	Miller, Freedson, & Kline, 1994
Baecke questionnaire[c]				.07	
Godin-Shephard questionnaire[c]				.30	
Minnesota leisure time METs				.36*	Jacobs et al., 1993
Caltrac accelerometer,[b] kcal				.12	
Caltrac accelerometer,[b] METs (14 two-day sessions)	73 males and females			.33*	
$\dot{V}O_2max$				.30*	
Submaximal (treadmill)	64 males and females			.32*	

[a]Scored in the same way as the 7-day questionnaire/interview.
[b]To be discussed in chapter 7.
[c]To be discussed later in the chapter.
*Statistically significant, $p < .05$.

undertook an exercise program, the questionnaire/interview score reflected only to a limited extent the change in fitness; the coefficient of correlation between the change in energy expenditures and the change in $\dot{V}O_2max$ was only .33 (Blair et al., 1985).

Baecke Questionnaire

A short, self-administered questionnaire for the assessment of habitual levels of physical activity was tested in a Dutch population (Baecke, Burema, & Frÿtters, 1982). Factor analysis identified three components of physical activity: at work, sport, and leisure excluding sport. Subjects responded to a 5-point scale with descriptors ranging from never (1 point) to always (5 points). A sport activity index was calculated as the sum of intensity multiplied by time engaged, which in turn was multiplied by proportion of the year participated. Similar indices were calculated for work and leisure. The questionnaire and formulas are reproduced in Appendix G.

Validity. Table 6.7 summarizes validation data for the Baecke questionnaire, although again the criteria employed leave much to be desired.

Reliability. Among 277 adult males and females, Baecke, Burema, and Frÿtters (1982) reported a test-retest correlation coefficient of .88 (work), .81 (sport), and .74 (leisure time). Three months intervened between the two administrations of the questionnaire. Jacobs et al. (1993) in 78 adults with 1 week intervening found test-retest coefficients of .93 for the total score, .78 for work, .90 for sports, and .86 for other leisure.

Health Insurance Plan of New York Questionnaire

A simple, self-administered questionnaire was developed for a study of coronary heart disease in New York City (Shapiro, Weinblatt, Frank, Sager, & Densen, 1963). The indexes of values are presented in Appendix H.

Validity. The validity of the Health Insurance Plan of New York Questionnaire rests primarily on the relationship of the physical activity assessments to coronary heart disease (Cassel et al., 1971; Frank, Weinblatt, Shapiro, & Sager, 1966a, 1966b; Hennekens, Rosner, Jesse, Drolette, & Speizer, 1977; Shapiro, Weinblatt, Frank, & Sager, 1965). The limited additional validation is shown in Table 6.8.

Reliability. When the original self-administered mail questionnaire was developed, 66 subjects were interviewed. The respondents were classified by the mail questionnaire and interview into three physical activity categories (light, intermediate, and heavy) using the point system shown in Appendix H. Agreement between the mail questionnaire and interview was 60.5% (Shapiro et al., 1965). More recently a test-retest (with 1 month intervening) correlation coefficient of .86 resulted with 78 adult male subjects (Jacobs et al., 1993).

British Civil Servant Questionnaire

An interview/questionnaire developed by Yasin (1967) and Yasin, Alderson, Marr, Pattison, and Morris (1967) has been used in studies of heart disease in British male executive civil servants. Only leisure time activities were recorded in the interview, which required about an hour. The respondents were asked to recall all activities lasting 5 min or more during the previous 2 days. Points are awarded per 5 min of reported activity, ranging from 1 point for sedentary activities such as reading to 5 points for sport activities such as tennis. The daily activity score is the sum of the points (maximum 60). This questionnaire/interview did not include occupational physical activity because it was designed for civil servants in sedentary occupations. It is not known whether this procedure is suited to other less-well-educated populations. It is also a time-consuming method, and the activity information covers only a brief period (one season of the year). The questionnaire form and an example from Andersen, Masironi, Rutenfranz, and Seliger (1978) are reproduced in Appendix I.

Validity. When this questionnaire was used, those rated less active were found to develop more coronary heart disease or higher risk factors associated with it than those rated active (Chave, Morris, Moss, & Semmence, 1978; Epstein, Miller, Stitt, & Morris, 1976; Morris, Chave, Adam, & Sirey, 1973; Morris, Everitt, Pollard, & Chave, 1980). These results, however, have not yet been duplicated by other investigators. In the original report (Yasin et al., 1967), a correlation coefficient of .27 was reported between the questionnaire/interview assessment of physical activity (8 days) and a 1-week record of weighted caloric intake in 32 adult males. The inactive group had a daily mean kilocalorie intake per kilogram of body weight of 36.9, the intermediately active group 40.9, and the most active group 41.9. The average point score on 4 work days was correlated with the average point score on 4 weekend days (r = .55) in 117 adult males. Similar results with regard to weighted caloric intake were reported by Marr, Gregory, Meade, Alderson, and Morris (1970) in 95 male office workers.

Framingham, Massachusetts, Questionnaire

The famous Framingham study included questions about physical activity in participant interviews (Appendix J contains the Framingham questionnaire from

Table 6.7 Baecke Activity Questionnaire Results Compared to Other Measures of Physical Activity

Other assessment method	Population	Correlation coefficient	Reference
Caltrac accelerometer[a] (4 days)	28 females	.53*	Mahoney & Freedson, 1990
Doubly labeled water, METs	28 males	.32*	Gretebeck et al., 1993
Caltrac accelerometer[a] (7 days) METs		.52*	
Caltrac accelerometer[a], kcal		.20*	
Large-Scale Integrated Motor Activity Monitor[a]	130 females (exercisers) 125 females (controls)	−0.11 to .20* .09 to .16*	Cauley et al., 1987
Paffenbarger questionnaire	130 females (exercisers) 125 females (controls)	.09 to .48* .14 to .36*	Cauley et al., 1987
Caloric intake	21 males	.38*	Albanes et al., 1990
Paffenbarger questionnaire		.56*	
7-Day recall questionnaire		.16	
Framingham questionnaire[b]		.57*	
Minnesota leisure time questionnaire		.36*	
Health Insurance Plan of New York questionnaire[b]		.78*	
Lipid Research Clinics questionnaire[b]		.68*	
Three 24-hr recall (kcal)	31 males and females, ages 63-80	.78*	Voorrips, Ravelli, Dongelmans, Deurenberg, & Van Staveren, 1991
Pedometer[a]		.72*	
Caltrac accelerometer[a], kcal (7 days)	7 males and 26 females	.32	Miller et al., 1994
7-Day recall questionnaire		.07	
Godin-Shephard questionnaire[b]		.61*	
$\dot{V}O_2$max	64 males and females	.54*	Jacobs et al., 1993
Submaximal (treadmill)		.51*	
Minnesota leisure time questionnaire (14 administrations)	73 males and females	.37*	
Caltrac accelerometer[a], METs (14 two-day sessions)		.19*	
Caltrac accelerometer[a], kcal		−0.08	

[a]To be discussed in chapter 7.
[b]Discussed later in the chapter.
*Statistically significant, $p < .05$.

Dannenberg & Wilson, 1989). A physical activity index was calculated by summing the products of the hours at each level of occupational and leisure time activity multiplied by a weight factor based on the oxygen consumption required for that activity (Kannel & Sorlie, 1979). The oxygen consumptions and weight factors (in parentheses) used are as follows: basal level (sleep, rest), 0.25 $L \cdot min^{-1}$ (1.0); sedentary activities (standing), 0.28 $L \cdot min^{-1}$ (1.1); slight level of activity (walking), 0.41 $L \cdot min^{-1}$ (1.5); moderate activity, 0.60 $L \cdot min^{-1}$ (2.4); heavy activity, 1.25 $L \cdot min^{-1}$ (5.0) (Kannel & Sorlie, 1979).

Validity. The index correlated moderately (r = .33 to .75) with other questionnaires (Albanes, Conway, Taylor, Moe, & Judd, 1990). The index has been found to be correlated with coronary heart disease (Garcia-Palmieri, Costas, Cruz-Vidal, Sorlie, & Havlik, 1982; Kannel & Sorlie, 1979; Yano, Reed, & McGee, 1984).

Reliability. Correlation coefficients between two successive measures ranged from .30 to .59 with 2-1/2 to 3 years intervening (Garcia-Palmieri et al., 1982).

Table 6.8 Health Insurance Plan of New York Activity Questionnaire Results Compared to Other Measures of Physical Activity

Other assessment measures	Population	Correlation coefficient	Reference
Doubly labeled water METs	28 males	.30*	Gretebeck et al., 1993
Caltrac accelerometer[a] METs (7 days)		.30*	
Energy intake METs (7 days)		.31*	
Caloric intake, food weighed	21 males	.19	Albanes et al., 1990
Paffenbarger questionnaire		.53*	
7-Day recall METs		.40*	
Framingham questionnaire[b]		.75*	
Minnesota leisure time questionnaire		.52*	
Baecke questionnaire		.78*	
Lipid Research Clinics questionnaire[b]		.68*	
Tecumseh questionnaire	197 males	.49*	Buskirk et al., 1971
Treadmill endurance	175 males	.07	Leon et al., 1981
Minnesota leisure time questionnaire, METs (14 administrations)	73 males and females	.00	Jacobs et al., 1993
Caltrac accelerometer[a], METs (14 two-day sessions)		.14	
Caltrac accelerometer[a], kcal		.07	
$\dot{V}O_2max$		.07	
Submaximal (treadmill)	64 males and females	.23*	

[a]To be discussed in chapter 7.
[b]Discussed later in the chapter.
*Statistically significant, $p < .05$.

Other Questionnaire/Interviews

We discuss several other questionnaire/interviews in this section, but their treatment is less detailed because the literature contains limited evidence of their validity or reliability, or they may have been used in a specific population and the suitability for other groups may be questionable.

Göteborg, Sweden, Questionnaire. In another well-known study carried out in Göteborg, Sweden (Wilhelmsen, 1969), men were classified into one of four groups on the basis of physical activity during work and into one of four groups on the basis of leisure time activity. Men judged active using this questionnaire experienced less coronary heart disease than did those judged more sedentary (Salonen, Puska, & Tuomilehto, 1982; Wilhelmsen & Tibblin, 1970), and the same was true of women (Lapidus & Bengtsson, 1986). Active men also had significantly higher $\dot{V}O_2max$ (Grimby, Wilhelmsen, Björntorp, Saltin, & Tibblin, 1971).

Lipid Research Clinics Questionnaire. The Lipid Research Clinics Coronary Prevention Trial utilized an interview in which the subjects were asked to respond to three or four questions about habitual physical activity (Gordon, Witztum, Hunninghake, Gates, & Glueck, 1983). Despite the simplicity of the assessment, the regular vigorous exercisers among 2,067 women and 2,319 men had lower heart rates at rest and during exercise than the more sedentary subjects (Haskell, Taylor, Wood, Schrott, & Heiss, 1980). These results were later confirmed by other investigators (Ainsworth, McNally, Jacobs, Cuptill, & Leon, 1989; Gordon et al., 1987; Reaven, McPhillips, Barrett-Connor, & Criqui, 1990). The physical activity classification resulting from the Lipid Research Clinics interview was correlated with the results of the Minnesota Leisure Time Activity Questionnaire results in 21 men (Albanes et al., 1990) and in 29 men and 61 women (Ainsworth et al., 1989).

Albanes et al. also reported that the results of the Lipid Research Clinics interview were significantly correlated with weighted caloric intake and the results of the Paffenbarger, 7-Day recall, Framingham, Baecke, and Health Insurance Plan of New York questionnaires. However, Jacobs et al. (1993) reported only low correlation with the Caltrac accelerometer (see chapter 7 for a discussion) and the average of 14 four-week activity histories with 73 adults. The Lipid Research Clinics Questionnaire correlated significantly (r = .49) with $\dot{V}O_2$max in most of these subjects. The same authors reported a test-retest coefficient of .93 using the 4-point scale. Ainsworth, Jacobs, and Leon (1993) studied two ways of scoring the Lipid Research Clinics Questionnaire in which subjects were classified into either two categories of habitual activity (2-point) or four categories (4-point). With about 1 month intervening, the test-retest coefficients were .85 and .88, respectively. The 4-point classification was slightly better in discriminating among subjects on the basis of physical fitness, body composition, and leisure activity.

Historical Physical Activity. Kriska et al. (Kriska et al., 1988; 1990; Kriska, LaPorte, & Knowler, 1992) presented a questionnaire to assess lifetime physical activity. Using this form, they were able to show a history of physical activity to be related to both bone area and density and favorable diabetes measurements. In 223 women, the activity reported in the most recent period in the historical physical activity survey, including walking, was weakly correlated with results of the Paffenbarger Questionnaire (.41) and the Large-Scale Integrated Motor Activity Monitor (see chapter 7) day counts (.12) and evening counts (.01). Twenty-three women completed the historical physical activity survey twice, 3 months apart. Test-retest scores agreed moderately well (Kappa values .39 to .47 and correlation coefficients .69 to .85).

National Surveys. A number of physical activity questionnaires have been developed for national surveys. These are usually simple to administer and sometimes are suitable for telephone surveys. They often require less than 10 min for response. Examples are the National Survey of Personal Health Practices in the United States (National Center for Health Statistics, 1980, 1981a, 1981b), Centers for Disease Control Behavioral Risk Surveillance (Bradstock et al., 1984; Yeager, Macera, & Merritt, 1991; National Health Interview Survey, National Center for Health Statistics, 1978; Weiss et al., 1990; Stephens & Craig, 1989), The Gallup Poll (1984), Health and Welfare Canada and Statistics Canada (1981), the Canadian Fitness Survey, modeled after the Minnesota Leisure Time Activity Questionnaire (Canada Fitness Survey, 1983; Stephens, 1983; Stephens & Craig, 1989; Stephens, Craig, & Ferris, 1986), Physical Activity of Finnish Adults (Voulle, Telama, & Laakso, 1986), President's Council on Physical Fitness and Sports (1974), The Perrier Survey (1979), and The Miller Lite Report on American Attitudes Toward Sport (1983). These questionnaires and interviews have not been validated, but they are thought to estimate the exercise habits of a nation. The physical activity portion of the 1991 U.S. National Health Survey form is reproduced as Appendix K because extensive data have been accumulated using it. A 24-hr recall leisure time questionnaire repeated four times during the year was used to assess the extent of physical activity in American adults by Brooks (1987a, 1987b, 1988a, 1988b).

Physical Activity, Health, and Physical Fitness. Several other activity questionnaires should be mentioned because they have been used in health and lifestyle studies, although neither the validity nor the reliability has been studied. Frisch and colleagues at Harvard University have published a series of papers in which female college graduates were queried by mail questionnaire about their former and present physical activity (Frisch et al., 1985, 1987; Frisch, Wyshak, Albright, Albright, & Schiff, 1986; Wyshak, Frisch, Albright, Albright, & Schiff, 1986, 1987). In a prospective study in Alameda County, California, physical activity by self-administered questionnaire was related to present and future health (Breslow & Enstrom, 1980; Wiley & Camacho, 1980; Wingard, Berkman, & Brand, 1982). Present physical activity has also been studied by mail questionnaire by investigators at the Aerobics Institute in Dallas, Texas, who reported significant correlation of physical activity with maximal treadmill performance (Blair, Kannel, Kohl, Goodyear, & Wilson, 1989; Kohl, Blair, Paffenbarger, Macera, & Kronenfeld, 1988; Kohl, Harris, & Blair, 1993).

Other physical activity questionnaires and interviews have been used to study the relationship of habitual exercise, or lack of it, to disease and disease-related risk factors. These include studies of adult men in The Netherlands (Magnus, Matroos, & Strackee, 1979); 2,014 working Norwegian males (Mundal, Erikssen, & Rodahl, 1987); about 15,000 Norwegian men aged 40 to 49 (Holme, Helgeland, Hjermann, Leren, & Lund-Larsen, 1981); 15,171 men in Ireland (Hickey, Mulcahy, Bourke, Graham, & Wilson-Davis, 1975); 370 middle-aged Danish men (Gyntelberg & Ohlsen, 1973); 2,109 Belgian males aged 40 to 55 (Sobolski et al., 1987); 77 males and 41 females in Great Britain (Lamb & Brodie, 1991); 8,171 men aged 45 to 64 in Puerto Rico (Costas, Garcia-Palmieri, Nazario, & Sorlie, 1978); 1,000 aviators of the U.S. Navy (York, Mitchell, & Graybiel, 1986); 1,004 men and women in the U.S. (Schechtman, Barzilai, Rost, & Fisher, 1991); 83 females aged 30 to 85 in the U.S. (Stillman, Lohman, Slaughter, & Massey, 1986); and 401 adults in the U.S. (Ross & Hayes, 1988). Some of these and other activity questionnaires not discussed earlier in the chapter have also been found to be predictive of endurance capacity (Bouchard et al., 1983; Gionet & Godin, 1989; Hopkins, Wilson, & Russell, 1991; Lamb & Brodie, 1991; Mundal et al.,

1987; Owen, Sedgwick, & Davies, 1988; Schechtman et al., 1991; Vogel, 1989).

The Saltin and Grimby (1968) questionnaire for occupation and leisure time developed for a Swedish study has been used with 43 middle-aged women (Wilbur, Miller, Dan, & Holm, 1989). The occupational part correlated significantly (r = .29) with the Tecumseh occupational questionnaire and with activity monitored by a beeper (r = .29). The leisure part did not correlate significantly with the Minnesota leisure time questionnaire but did correlate significantly with beeper monitoring (r = .44). A one-question questionnaire (How often do you engage in exercise which causes you to sweat and become breathless?) was used in Australia (Owen, Sedgwick, & Davies, 1988). The response was moderately related to an estimate of $\dot{V}O_2max$. Although many of these correlation coefficients are statistically significant, one must bear in mind that a coefficient of .3 accounts for less than 10% of the variance.

Elderly Adults

Physical activity is believed to play an important role in the maintenance of health and effective function in older people. Evidence from the Alameda County study showed that, among the elderly, participation in leisure time physical activity was associated with a decreased 17-year follow-up mortality risk independent of age, socioeconomic status, health status, smoking, relative weight, and alcohol consumption (Kaplan, Seeman, Cohen, Knudsen, & Guralnik, 1987). A report by Mor et al. (1989) showed that in a cohort aged 70 to 74, those who did not report participation in regular exercise or walking a mile without resting were 1.5 times more likely to suffer a decline in functional status over 2 years after controlling for medical conditions and demographic factors. In Dutch men 65 years and older, physical activity was significantly related to some coronary heart disease risk factors, although the relationship was weak (Caspersen, Bloemberg, Saris, Merritt, &Kromhout, 1991). Physical activity has also been associated with a decreased risk for falls and fractures and with increased bone density (Campbell, Borrie, & Spears, 1989; Dalsky et al., 1988; Sorock et al., 1988).

The principles and limitations of questionnaires and interviews discussed in the previous section on young and middle-aged adults generally apply when assessing physical activity in the elderly. Education and socioeconomic level should be considered when designing a questionnaire or interview for the elderly, as for other populations. In older people, these two factors have been related to errors in reporting physical activity (Washburn, Jette, & Janney, 1990).

Most elderly subjects are retired, so leisure time activity may become all-important, particularly household tasks, gardening, and walking. Only a small percentage of older people participate in moderate or strenuous physical activity (Blair et al., 1985; Dannenberg, Keller, Wilson, & Castelli, 1989). Nevertheless, some elderly still engage in sports. Allowance must be made for hearing and visual limitations when surveying older subjects. An oral questionnaire administered during a personal interview may be the best approach. A number of subjects in this age group may also have physical or medical conditions that restrict physical activity considerably. Information about these conditions should be sought in response to the questionnaire. Although physical activity may change during the year among people of any age, seasonal fluctuations are greater in the elderly (Uitenbroek, 1993). Another problem is encountered if physical activity data are to be converted to energy expenditure values in the elderly. Most data on the energy cost of activities are based on testing with young adults. Most elderly people probably perform activities at lower intensities (lower energy costs) than young, healthy subjects, and some adjustments might be necessary.

Baecke Questionnaire Modification

The Baecke self-administered questionnaire discussed earlier in the chapter was modified for use with older people (Voorrips, Ravelli, Dongelmans, Deurenberg, & Van Staveren, 1991). The modified form (reproduced in Appendix L) was designed to encompass the previous year's activities and to be completed by an interviewer rather than self-administered. Because household activities become more important for the elderly, additional questions about these activities were included. A modified scoring system was also developed (see Appendix L).

Validity. The validity of the questionnaire/interview was evaluated by having 31 different men and women in the same age group complete a 24-hr recall questionnaire on 3 days (including 1 weekend day). Activities were recorded in 10-min intervals. Thirty of these subjects also wore pedometers for 3 days (1 weekend day). The Spearman coefficient expressing the relationship between the modified Baecke questionnaire/interview and the 24-hr recall score was .78, with 71% of the participants classified in the same tertile and one grossly misclassified. The Spearman coefficient for a Baecke-pedometer comparison was .72, with 67% classified in the same tertile and no gross misclassifications.

Reliability. To test the reliability of the questionnaire/interview, the assessment was repeated after 20 days in

29 men and women, aged 60 to 83 years. The mean scores of the two administrations did not differ significantly. The Spearman rank order coefficient of correlation for the two scores was .89. Seventy-two percent of the subjects were classified in the same tertile on the two administrations, and gross misclassification (two scores in the opposite tertiles) did not occur.

Zutphen Questionnaire

A physical activity questionnaire for retired men was made available by J.N. Morris of the London School of Hygiene and Tropical Medicine. This was modified to be self-administered for use with 863 men aged 65 to 84 living in Zutphen, The Netherlands (Caspersen et al., 1991). It is reproduced in Appendix M, together with the intensity codes that were used. The score was expressed as kilocalories per kilogram of body mass per day. Time estimates were converted to minutes per week for each type of activity. The kilocalorie score was calculated by converting the minutes per week for each activity into hours per day. The result was then multiplied by the intensity code (see Appendix M) expressed as kilocalories per kilogram body weight times hour to reflect multiples of resting metabolism. Also tallied were the total minutes per week spent in light (<2 kcal/kg · hr), moderate (≥2 kcal/kg · hr but <4 kcal/kg · hr), and heavy ≥4 kcal/kg · hr) physical activity. The questionnaire was used successfully with a large sample of elderly men.

Validity. Significant relationships were found between the physical activity score and some coronary risk factors (Caspersen et al., 1991).

Recently the questionnaire was also validated using a sample of the same population, then 5 years older (Saris et al., 1993). In this select sample, energy expenditure was measured by the doubly labeled water (DLW) method and resting metabolism was determined by measuring the oxygen consumed. The ratio of total energy expended (by DLW method) to resting metabolism was correlated with the physical activity questionnaire score. The resulting statistically significant ($p < 0.05$) correlation was .62.

Reliability. The reliability of the questionnaire was studied with 21 subjects selected on the basis of their activity scores: 8 below the 10th percentile, 5 in the 45th to 55th percentile range, and 8 above the 90th percentile (Saris et al., 1993). The test-retest coefficient was .93 with 3 months intervening. The more active the subjects, the less consistent were their activity scores.

Yale Physical Activity Survey

The Yale Physical Activity Survey (YPAS; DiPietro, Caspersen, Ostfeld, & Nadel, 1993) is an interview-administered questionnaire developed for assessing habitual physical activity in elderly subjects; it takes about 20 min to complete (see Appendix N). Several indices are calculated from the subjects' responses, including the following:

- Total time. The times for each checklist activity are summed and expressed as hours per week.
- Energy expenditure. Times for each checklist activity are multiplied by appropriate intensity codes and then summed. The intensity codes are defined in terms of a resting metabolic rate and hence are independent of body weight. The index is expressed as kcal · wk^{-1}.
- Activity dimensions. A summary index is calculated by multiplying a frequency score by a duration score for each of five classes of activities. The five classes of activities are assigned a weighting factor depending on the strenuousness of the activity. The summary index is the sum of the five individual indices.

Validity. YPAS validity was evaluated by comparing the results to scores on a Caltrac accelerometer (to be discussed in chapter 7) worn by each subject for 2.5 weekdays, $\dot{V}O_2$max estimated from a submaximal treadmill test, and resting diastolic blood pressure, body mass index, and percent body fat. Twenty-five subjects, 14 men and 11 women, comprised the validation study. The results are shown in Table 6.9. The validity is not impressive, although, as we noted before, these criteria leave much to be desired.

Reliability. Seventy-six subjects (20 men and 56 women) aged 60 to 86, completed the questionnaire/interview twice with 2 weeks intervening. The test-retest comparison resulted in only fair reproducibility. The second administration produced lower scores than the first in the three indices (only in the first index was the difference statistically significant at p of 0.05). Test-retest Pearson product moment correlation coefficients for the three indices were .57, .58, and .65, respectively.

Physical Activity Scale for the Elderly (PASE)

A questionnaire to assess physical activity was developed by a group at the New England Research Institute (Washburn, Smith, Jette, & Janney, 1993). A sample of 36 males and females aged 65 to 85+ was used to refine a draft of the questionnaire. The revised questionnaire was used with 277 men and women who participated in both the questionnaire and a home visit and

Table 6.9 Spearman Rank Order Correlation Coefficients Between Yale Physical Activity Survey (YPAS) Indices and Validation Criteria (N = 25)

YPAS indices	Estimated $\dot{V}O_2max$	Resting diastolic blood pressure	Body mass	Percent body fat	Caltrac accelerometer[a]
Total time	0.03	–0.35	–0.23	–0.04	0.08
Energy expenditure	0.20	–0.47*	–0.15	–0.07	0.14
Summary index	0.58*	–0.21	–0.01	–0.43*	0.37

Note. From DiPietro, Caspersen, Ostfeld, & Nadel, 1993.
[a]To be discussed in chapter 7.
*Statistically significant, $p < 0.05$.

119 who participated in the questionnaire phase only. All subjects were aged 65 or older, with mean ages of both groups being 73 years. Some in the group were interviewed by phone, and the rest completed mailed questionnaires. Home visits included measurements of resting heart rate, blood pressure, stature, weight, grip strength, balance, knee extensor strength, and health status by the Sickness Impact Profile. Activity was monitored for three days by the Caltrac accelerometer (to be discussed in chapter 7). A diary was also maintained during these 3 days, and a habitual physical activity rating was completed by the subject. A sample of the subjects completed the questionnaire a second time, 3 to 7 weeks later. The scoring system was developed through a multiple regression of the questionnaire items against the criterion consisting of the 3-day diary, 3-day Caltrac monitoring, and self-assessment of habitual activity.

Test-retest correlation coefficients were .68 for the telephone survey, .84 for the mail survey, and .75 for the combined groups. PASE scores were positively associated with grip strength (r = .37), static balance (r = .33), and leg strength (r = .25) and were negatively correlated with heart rate (r = –.13), age (r = –.34), and Sickness Impact Profile (r = –.42). All of the coefficients were statistically significant at $p < .05$.

The PASE manual and copies of the questionnaire are sold by the New England Research Institute, 9 Galen St., Watertown, MA, USA 02172.

Children

Estimating physical activity or energy expenditure by questionnaire or interview in children generally requires a different procedure or survey instrument (or both) than when only adults are involved. Moreover, the potential errors in recall are very likely greater. In fact, most experienced investigators agree that the questionnaire approach is not appropriate with children younger than about age 10. With these younger children, other methods (observation, instrument monitoring, etc.) are often used instead, or the questionnaire/interview is completed by a parent, teacher, or other adult.

The interview/questionnaire procedure differs between children and adults because the source of activity to some extent differs. Young children are not employed, so there is no occupational activity. In the case of an older child who has a job, an adult occupational questionnaire/interview must be used unless the employment is for only brief periods. The questionnaire/interview with children must contain inquiries about participation in physical education classes, recess periods, organized sports (intramural, school, or club), transportation to and from school, and recreational activities after school and during weekends. As with adults, information on intensity, frequency, and duration of activities should be sought. The language must be suitable to the age of the child. It is also helpful for recall to list activities for the child to check. Also, if activities are to be classified by intensity (METs), the investigator, not the child, parent, or teacher, should assign the classification, although the child could be asked questions about sweating, heart rate, and the like. It is also important to provide a time frame, such as what the child did on the way to school, in the morning at school, in the afternoon at school, returning home from school, after school before dinner, and after dinner.

Recall of activities by children can be inaccurate. Wallace, McKenzie, and Nader (1985) reported that 11- to 13-year-old children recalled only 46% of their activities during the previous 7 days and 55% to 65% during the previous 24 hr as documented by three adult observers. Baranowski et al. (1984) reported similar results in third to sixth graders.

In general, children cannot estimate the duration of an activity very well. In the mind of a child, intensity and joy are often important criteria for the duration of an activity. By using certain time points, it is possible to improve accuracy, so any questionnaire adapted for use with children should include specified time periods.

As indicated in the previous two sections, the validation and reproducibility of adult questionnaires/interviews of physical activity leave much to be desired. The situation is even worse with this approach in children, which has received less critical study. Often in reports physical activities of boys and girls are tabulated from questionnaire results, but the actual questionnaire form is not provided and no data on validity or reliability are given. Examples include studies of Finnish children aged 12 to 18 years (Laakso & Telama, 1981), eighth-grade Swedish children (Engström, 1980); Canadian girls aged 9 to 13 (Moisan, Meyer, & Gingras, 1991); a national survey of children aged 6 to 18 years by the U.S. Public Health Service (Ross, 1989); and the Canadian national fitness survey for children aged 10 and older (Stephens & Craig, 1989).

Netherlands Health Education Project Questionnaire

One approach to assessing physical activity in children is exemplified by the Physical Activity Score (see Appendix O) used by Saris, Doesburg, Lemmens, and Reingis (1974), which was especially useful when this questionnaire was completed by the teacher or parents. The questions are not meant to be answered with involvement in specific activities; rather they are carefully selected items about the child's behavior in everyday situations at home or at school. An evaluation of this approach resulted in significant correlation (r = −.81, −.86) with observation and pedometer and actometer scores.

Northern Ireland Questionnaire

Riddoch (1990) used a 7-day recall and supplementary questionnaire to determine the activity levels of about 3,200 children aged 11 to 18 in Northern Ireland. Although neither validity or reliability is reported, the questionnaire appears to be carefully constructed and was used successfully in a large study.

Amsterdam Growth Study Questionnaire

The questionnaire used in the Amsterdam growth study (see Appendix P) was described by Verschuur and Kemper (1987). The period covered by the questionnaire/interview was the previous 3 months. Activities with a rating below 4 METs were not recorded during the interview or included in the score. The other activities were classified as light, medium-heavy, and heavy. For purposes of scoring, the METs assigned to these categories were 5.5, 8.5, and 11.5, respectively. The interview assessed the average weekly time spent doing activities in each of the three categories with a minimum of 5 min. The score was expressed as METs per week, derived by multiplying the average time spent per week in each category by the respective METs for that category. The scores of the three levels were added to arrive at a score of total METs per week.

The relationship of the results of the Amsterdam Growth Study Questionnaire and other measures of physical activity are shown in Table 6.10.

Adult Questionnaires Used With Children

The 7-day recall questionnaire discussed in the section on young and middle-aged adults has been used or adapted for use with children. The validation data are summarized in Table 6.11 and the reliability data in Table 6.12. The Minnesota Leisure Time Activity Questionnaire was used with children by LaPorte et al. (1982). With 22 boys aged 12 to 14, they reported no correlation (r = .02) when compared with monitoring with a Large-Scale Integrated Motor Activity Monitor (discussed in chapter 7).

Relationship of Physical Activity to Health in Children

Just as in adult studies, some activity questionnaires/interviews with children have been shown to be related to risk factors of disease, particularly cardiovascular disease. Some investigators have called this *indirect validation* or *face validity*. Studies of this kind in children include the report on 272 sixth-grade Italian children by Strazzullo et al. (1988), who modified the questionnaire of Saltin and Grimby (1968). Viikari and other Finnish co-workers (1984) had parents of 189 three-year-olds and parents of 237 twelve-year-olds or the children themselves complete activity questionnaires.

In another report of Finnish children (174 eight-year-old boys), Wanne et al. (1983) provided their simple questionnaire, which was filled out by parents. They also reported correlations of the results of the questionnaire with risk factors. In still another study of Finnish children, Marti and Vaartiainen (1989) used both a self-administered questionnaire/interview and a parent questionnaire to study risk factors and activity among 15-year-olds. Verschuur, Ritmeester, Kemper, and Storm-van Essen (1987) and Kemper, Suel, Verschuur, and Storm-van Essen (1990) compared coronary heart disease risk factors with questionnaire/interview assessment of physical activity in Dutch children. Durant, Linder, Harkness, and Gray (1983) and Durant, Linder,

Table 6.10 Amsterdam Growth Study Questionnaire: Validity

Questionnaire/interview	Population	Criterion	Correlation	Reference
Questionnaire/interview total activity time	52 boys			Verschuur, Kemper, & Storm-van Essen, 1987
	Ages 12-13	Endurance run	$r = .25$*	
	Ages 16-17		$r = .26$*	
	Ages 12-13	Bicycle ergometer	r not significant	
	Ages 16-17		$r = .24$*	
Questionnaire/interview sports activity time	52 boys			
	Ages 12-13	Endurance run	$r = .29$*	
	Ages 16-17		$r = .52$*	
	Ages 12-13	Bicycle ergometer	r not significant	
	Ages 16-17		r not significant	
Questionnaire/interview	195 boys, 215 girls, ages 13-14	$\dot{V}O_2max$	Significant correlation in boys but not in girls	Verschuur & Kemper, 1985a, 1985b
Questionnaire/interview	102 boys, 131 girls, ages 13-17	Heart rate monitoring[a]	$r = .16$ to $.18$	Kemper, 1992
		Pedometer[b]	$r = .17$	

[a]To be discussed in chapter 8.
[b]To be discussed in chapter 7.
*Statistically significant, $p < .05$.

Table 6.11 7-Day Recall Questionnaire Adapted for Children: Validity

Population	Criterion	Correlation	Reference
25 boys and girls, mean age 12	$\dot{V}O_2max$	$r = .67$*	Schmücker, Rigauer, Hinrichs, & Trawinski, 1984
39 boys, 58 girls, ages 10-15	Submaximal bicycle ergometer	$r = .40$* (boys) $r = .23$ (girls)	Suter & Hawes, 1993
11 boys, mean age 12.5	Direct observation	Fair agreement	Wallace et al., 1985
69 boys and girls, 4th grade	Caltrac accelerometer[a]	$r = .36$* $r = .36$* $r = .34$* $r = .11$	Roby, Sallis, Kolody, Condon, & Goggin, 1992
102 boys and girls, 5th, 8th, and 11th grades	Heart rate monitoring[b]	$r = .47$ to $.81$ for various age groups	Sallis, Buono, Roby, Micale, & Nelson, 1993; Sallis et al., 1991

[a]To be discussed in chapter 7.
[b]To be discussed in chapter 8.
*Statistically significant, $p < .05$.

and Mahoney (1983) correlated physical activity and risk factors in 149 children aged 7 to 17 in the United States using a questionnaire/interview. Sallis et al. (1988), using a 7-day recall/interview, did the same with 290 Anglo and Mexican children (mean age 12 years). In this last study, a self-administered questionnaire produced the same results. Suter and Hawes (1993) also used a 7-day recall questionnaire to study the relationship of physical activity and heart disease risk factors in 97 children in the U.S. Using a weekly questionnaire on history of physical activity completed by the child or parent, Ludvigsson (1980) was able to show a correlation between regular physical activity and metabolic control in Swedish diabetic children. Many of

Table 6.12 7-Day Recall Questionnaire Adapted for Children: Reliability

Population	Period between observations	Coefficient of correlation	Reference
290 boys and girls, mean age 12	Same day, different interviewers	.78	Sallis, Patterson, Buono, Atkins, & Nader, 1988
102 boys and girls, 5th, 8th, and 11th grades	—	.47 to .81 for different ages	Sallis, 1991; Sallis et al., 1993
690 boys and girls, grade 7-9	2 weeks	.84	Godin & Shephard, 1984
102 boys and girls, 5th, 8th, and 11th grades	2 weeks	.69 to .96 for different ages	Sallis, 1991
69 boys and girls, 4th grade	3 days	.55 (7 days) .66 (24 hr) .75 (7 days, Var I) .62 (7 days, Var II)	Roby et al., 1992

these studies have been reviewed elsewhere (Montoye, 1985, 1986).

Additional Questionnaires/Interviews Used With Children

Other questionnaires/interviews that have been used with children are listed in Tables 6.13 through 6.16. Correlations of the results with direct observation are shown in Table 6.13, with activity monitoring in Table 6.14, and with physical fitness in Table 6.15. Test-retest comparisons are given in Table 6.16.

SUMMARY

Although the validity and test-retest reproducibility of questionnaires/interviews concerning physical activity have not been adequately studied, much useful information can be obtained by their use. This approach is the only feasible one for epidemiologic investigations. Despite the limitations of the method, the results are often correlated with longevity and morbidity in populations. In selecting a particular questionnaire/interview, it is necessary to determine the purposes for using the instrument, time and financial constraints, and the sex, age, and socioeconomic characteristics of the population. The questionnaire methods appear to be inappropriate for children under 10 to 12 years old.

In young and middle-aged subjects who are working, it is necessary to assess both occupational and leisure time physical activities. The questionnaire method will probably not provide accurate individual energy expenditures, but it should be possible to group people into three to five categories on the basis of habitual physical activity. Strenuous physical activity appears to be recalled with greater accuracy than mild or moderate activity, and recall of recent activities is more accurate than those done at an earlier time. Both weekends and weekdays should be covered as well as the various seasons of the year, because the activities of some individuals vary depending on the day of the week or the season. Few questionnaires or interviews are designed specifically for the elderly, and those that are available need further study with regard to validity and reproducibility. Alternatively, new questionnaires/interviews might be developed that are specific to the older age group. The following guidelines are suggested for doing this:

- Activity participation is difficult for older persons to recall accurately. Recall is enhanced if the time frame over which activity is assessed is short, such as the previous week.
- The questionnaire should emphasize the domains of physical activity most likely to be performed by older people. For example, specific information should be obtained about walking, light moderate housework, outdoor work, and so forth.
- Specific questions should be short and focused on one type of activity to eliminate problems associated with accurately discriminating between activities. For example, it is difficult to respond to a general question about how much walking one does. Instead, the questionnaire should ask specifically about walking for errands, for exercise, with

Table 6.13 Relationship of Additional Physical Activity Questionnaire/Interviews With Direct Observation in Children

Questionnaire/interview	Population	Correlation	Reference
Questionnaire by teacher	11 boys and girls, ages 4-6	Significantly related	Saris & Binkhorst, 1977
Questionnaire—physical education class	44 third-grade children	86% agreement	Simons-Morton et al., 1990
24-hr recall questionnaire	24 third- to sixth-grade children	70% to 80% agreement	Baranowski et al., 1984
Teacher rating Parent rating	21 boys and girls, ages 3-5, 20-min observation at school (video recording)	Not related	Noland, Danner, Dewalt, McFadden, & Kotchen, 1990
	8 boys and girls, ages 3-5, 6-hr observation at home (video recording)	Not related	

Table 6.14 Relationship of Physical Activity Assessed by Other Questionnaire/Interviews With Movement or Heart Rate Monitoring in Children

Questionnaire/interview	Population	Monitoring method	Correlation	Reference
Questionnaire by supervisor	23 boys and 25 girls, mean age 2.5	Monitor at school	$r = .41$ to $.60$	Halverson & Waldrup, 1973
Questionnaire/interview	71 boys, ages 12-13	Pedometer[a]	$r = .50$*	Kemper & Verschuur, 1974
24-hr recall questionnaire/ interview	20 boys and 15 girls, ages 8-13	Caltrac accelerometer	$r = .45$*	Sallis, Buono, Roby, Carlson, & Nelson, 1990
		Heart rate monitoring[b]	$r = .38$*	

[a]To be discussed in chapter 7.
[b]To be discussed in chapter 8.
*Statistically significant, $p < .05$.

the dog, around the house, and so on to elicit more accurate responses.

- An age-specific approach should also include a categorical response format that will focus recall of salient activities and the time spent in them. A format for days per week and minutes per time of activity participation may provide more meaningful information than an open-ended response format and may be more easily completed by respondents.
- Special attention should be given to questions relating to physical conditions that can limit daily physical activity.

It is difficult and generally not appropriate to try to estimate habitual physical activity in young children by questionnaire or interview. For young children, other techniques such as observation, movement counters, and heart rate recording must be considered first. If interview and questionnaire are the only alternatives, the possible errors must be reduced. Questionnaires and interviews should be focused on the daily routines, such as transportation to and from school, organized sports, and the like. Anther approach is to ask a parent, teacher, or supervisor about a child's attitudes toward physical activity (favorite play activities) in order to rank children according to their habitual physical activity.

Because most energy cost tables of physical activity are based on adult data, substantial errors in estimating energy cost are likely if these tables are used with the elderly or children. A thorough discussion of this problem, particularly as it applies to children, and a correction factor for walking and running may be found in the paper by Sallis, Buono, and Freedson (1991). The use of METs minimizes this error.

Table 6.17 presents a number of characteristics of some questionnaires/interviews as a guide to aid in selecting a particular instrument. Further discussion of physical activity questionnaires/interviews is available in the references by Washburn and Montoye (1986) and Ainsworth, Montoye, and Leon (1994).

Table 6.15 Relationship of Physical Activity by Other Questionnaire/Interviews With Physical Fitness in Children

Questionnaire/interview	Population	Fitness measurement	Correlation	Reference
Questionnaire by teacher and parent	2,352 boys and girls, mean age 4	Endurance performance	r = .00 to .22 for various tiems	Pate, Dowda, & Ross, 1990
Questionnaire by parent	171 boys and girls, ages 4-6	Treadmill performance	Significantly related	Saris, Binkhorst, Cramwinckel, Vander Veen-Hezemans, & Van Waesberghe, 1979
Questionnaire by teacher			Significantly related	
Questionnaire by parent or child			Not related	
Questionnaire by parent	54 boys and girls, ages 8-12	Treadmill performance	Significantly related	Saris et al., 1979
Questionnaire by teacher			Significantly related	
Questionnaire by parent or child			Not related	
Questionnaire, usual activity	413 boys and 372 girls, ages 10-14	$\dot{V}O_2max$	r = .12 boys r = .13 girls	Tell & Vellar, 1988
Questionnaire by supervisor	125 boys and 89 girls, ages 6-18	$\dot{V}O_2max$	Significantly related	Murphy, Alpert, Christman, & Willey, 1988
Questionnaire/interview	63 boys and girls, age 8 68 boys & girls, age 13	$\dot{V}O_2max$	r = .41* to .48*, various age/sex groups	Sunnergårdh & Bratteby, 1987
Self-report of the past year	608 boys and 567 girls, ages 12-16	Endurance run time	r = –.11* (boys) r = –.21* (girls)	Aaron et al., 1993

*Statistically significant, $p < .05$.

Table 6.16 Summary of Test–Retest Results With Other Activity Questionnaires and Interviews With Children

Type of questionnaire/interview	Population	Period between administrations	Coefficient of correlation	Reference
Interview, usual activity	50 boys, ages 11-17	8 weeks	.70	Linder, Durant, & Mahoney, 1983
Rating by supervisor	58 boys and girls, mean age 2.5	—	.88 interrater	Halverson & Waltrop, 1973
24-hr self-report	12 children, mean age 11	12 hr	.90	Janz, Phillips, & Mahoney, 1992
Godin-Shephard (1985) self-report	36 fifth-graders 36 eighth-graders 30 eleventh-graders	2 weeks	.81	Sallis et al., 1993
Self-report for the past year	1039 boys and girls, ages 12-16	1 year	.55	Aaron et al., 1993

Table 6.17 Characteristics of Particular Questionnaires/Interviews

Questionnaire/interview	Type of administration	Time frame	Approximate time to administer	Type of activity	Measurement scale
Young and middle-aged adults					
Tecumseh occupational	Interview	1 year	30 min	Occupational	Average METs
Minnesota leisure time	Interview	1 year	45 min	Leisure time	METs activity
Paffenbarger/Harvard	Self-administered	Usual activity	15 min	Occupational and leisure time	Index
Five-city/seven-day	Interview	1 week	<15 min	Occupational and leisure time	Kcal score
Baecke	Self-administered	Usual activity	15 min	Occupational and leisure time	Kcal score
Health Insurance Plan of New York	Self-administered or interview	Usual activity	15 min	Occupational and leisure time	3 indices (work, sport, leisure)
British civil service	Interview	Leisure time, specific days	1 hr	Leisure time	4 activity classes
Framingham	Interview	Usual acitivty	15 min	Occupational and leisure time	Activity index
Lipid Research Clinics	Interview	Usual activity	10 min	Occupational and leisure time (strenuous only)	Index (time and strenuousness)
Kriska	Self-administered	Life-long activity		Occupational and leisure time	4 activity classes
Elderly Adults					
Baecke	Self-administered	Usual activity	10 min	Leisure time	Activity index
Zutphen	Self-administered	Usual activity	15 min	Leisure time	kcal · kg^{-1} score
Yale Physical Activity Survey	Interview	Usual activity	20 min	Occupational and leisure time	Time, energy expenditure, and activity index
Children					
Netherlands Health Education Project	Self-administered, teacher or parent	Usual activity	10 min	Leisure time and school	Activity index
Northern Ireland	Self-administered	Usual activity and 1 week	30 min	Leisure time, school, and occupational	Activity index
Amsterdam Growth Study	Interview	3 months	20 min	Leisure time and school	METs activity index
Five-city/seven-day	Interview	1 week	<15 min	Leisure time and school	kcal score

REFERENCES

Aaron, D.J. Kriska, A.M., Dearwater, S.R., Anderson, R.L., Olsen, T.L., Cauley, J.A., & LaPorte, R.E. (1993). The epidemiology of leisure physical activity in an adolescent population. *Medicine and Science in Sports and Exercise,* **25,** 847-853.

Ainsworth, B.E., Jacobs, D.R., Jr., & Leon, A. (1990). Validity of assessment of physical activity using the college alumnus questionnaire [abstract]. *Medicine and Science in Sports and Exercise,* **22,** S79.

Ainsworth, B.E., Jacobs, D.R., Jr., & Leon, A.S. (1993). Validity and reliability of self-reported physical activity status: The Lipid Research Clinics Questionnaire. *Medicine and Science in Sports and Exercise,* **25,** 92-98.

Ainsworth, B.E., & Leon, A.S. (1991). Gender differences in self-reported physical activity [abstract]. *Medicine and Science in Sports and Exercise,* **23,** S105.

Ainsworth, B., McNally, C., Jacobs, D., Cuptill, Y., & Leon, A. (1989). Validity of self classification of physical activity status [abstract]. *Medicine and Science in Sports and Exercise,* **21,** S112.

Ainsworth, B.E., Montoye, H.J., & Leon, A.S. (1994). Methods of assessing physical activity during leisure and work. In C. Bouchard, R.J. Shephard, & T. Stephens, (Eds.), *Physical activity, fitness, and health* (pp. 146-159). Champaign, IL: Human Kinetics.

Albanes, D., Conway, J.M., Taylor, P.R., Moe, P.W., & Judd, J. (1990). Validation and comparison of eight physical activity questionnaires. *Epidemiology,* **1,** 65-71.

Andersen, K.L. Masironi, R., Rutenfranz, J., & Seliger, V. (1978). *Habitual physical activity and health.* Copenhagen: World Health Organization Regional Publication. (European Series No. 6)

Baecke, J.A.H., Burema, J., & Frÿtters, J.E.R. (1982). A short questionnaire for the measurement of habitual physical activity in epidemiological studies. *American Journal of Clinical Nutrition,* **36,** 932-942.

Baranowski, T., Dworkin, R.J., Cieslik, C.J., Hoods, P., Clearman, D.R., Ray, L., Dunn, J.K., & Nader, P.R. (1984). Reliability and validity of self-report of aerobic activity: Family health project. *Research Quarterly for Exercise and Sports,* **55,** 309-317.

Blair, S.N., Dowda, M., Pate, R.R., Kronenfeld, J., Howe, H.G., Jr., Parker, G., Blair, A., & Fridinger, F. (1991). Reliability of long-term recall of participation in physical activity by middle aged men and women. *American Journal of Epidemiology,* **133,** 266-275.

Blair, S.N., Haskell, W.L., Ho, P., Paffenbarger, R.S., Jr., Vranizan, K.M., Farquhar, J.W., & Wood, P.D. (1985). Assessment of habitual physical activity by a seven-day recall in a community survey and controlled experiments. *American Journal of Epidemiology,* **122,** 794-804.

Blair, S.N., Kannel, W.B., Kohl, H.W., Goodyear, N., & Wilson, P.W.F. (1989). Surrogate measures of physical activity and physical fitness. *American Journal of Epidemiology,* **129,** 1145-1156.

Borg, G.A.V. (1982). Psychological bases of physical exertion. *Medicine and Science in Sports and Exercise,* **14,** 377-381.

Bouchard, C., Tremblay, A., Leblanc, C., Lortie, G., Savard, R., & Theriault, G. (1983). A method to assess energy expenditure in children and adults. *American Journal of Clinical Nutrition,* **37,** 461-467.

Bradstock, M.K., Marks, J.S., Forman, M., Gentry, E.M., Hogelin, G.C., & Trowbridge, F.L. (1984). Behavioral Risk Factor Surveillance, 1981-1983. *Centers for Disease Control, Surveillance Summaries,* **33,** 155-455.

Breslow, L., & Enstrom, J.E. (1980). Persistence of health habits and their relationship to mortality. *Preventive Medicine,* **9,** 469-483.

Brooks, C. (1988a). A causal modeling analysis of sociodemographics and moderate to vigorous physical activity behavior of American adults. *Research Quarterly for Exercise and Sport,* **59,** 328-338.

Brooks, C. (1988b). Adult physical activity behavior: A trend analysis. *Journal of Clinical Epidemiology,* **41,** 385-391.

Brooks, C.M. (1987a). Adult participation in physical activities requiring moderate to high levels of energy expenditure. *Physician and Sports Medicine,* **15,** 119-132.

Brooks, C.M. (1987b). Leisure time physical activity assessment of American adults through an analysis of time diaries collected in 1981. *American Journal of Public Health,* **77,** 455-460.

Brouha, L. (1960). *Physiology in industry.* London: Pergamon Press.

Buskirk, E.R., Harris, D., Mendez, J., & Skinner, J. (1971). Comparison of two assessments of physical activity and a survey method for calorie intake. *American Journal of Clinical Nutrition,* **24,** 1119-1125.

Campbell, A.J., Borrie, M.J., & Spears, G.F. (1989). Risk factors for falls in a community-based prospective study of people 70 years and older. *Journal of Gerontology,* **44,** M112-117.

Canada Fitness Survey (1983). Fitness and lifestyle in Canada. Ottawa, ON: Author.

Caspersen, C.J., Bloemberg, B.P.M., Saris, W.H.M., Merritt, R.K., & Kromhout, D. (1991). The prevalence of selected physical activities and their relation with coronary heart disease risk factors in elderly men: The Zutphen Study, 1985. *American Journal of Epidemiology,* **133,** 1078-1092.

Cassel, J., Heyden, S., Bartel, A.G., Kaplan, B.H., Tyroler, H.A., Cornoni, J.C., & Hames, C.G. (1971). Occupation and physical activity and coronary heart disease. *Archives of Internal Medicine,* **128,** 920-928.

Cauley, J.A., LaPorte, R.E., Black-Sandler, R., Orchard, T.J., Slemenda, C.W., & Petrini, A.M. (1986). The relationship of physical activity to high density lipoprotein cholesterol in postmenopausal women. *Journal of Chronic Disease,* **39,** 687-697.

Cauley, J.A., LaPorte, R.E., Black-Sandler, R., Schramm, M.M., & Kriska, A.M. (1987). Comparison of methods to measure physical activity in postmenopausal women. *American Journal of Clinical Nutrition,* **45,** 14-22.

Chave, S.P.W., Morris, J.N., Moss, S., & Semmence, A.M. (1978). Vigorous exercise in leisure time and the death rate: A study of male civil servants. *Journal of Epidemiology and Community Health,* **32,** 239-243.

Christensen, E.H. (1953). *Symposium on fatigue.* London: Lewis.

Costas, R., Jr., Garcia-Palmieri, M.R., Nazario, E., & Sorlie, P.D. (1978). Relation of lipids, weight and physical activity to incidence of coronary heart disease: The Puerto Rico study. *American Journal of Cardiology,* **42,** 653-658.

Cunningham, D.A., Montoye, H.J., Metzner, H.L., & Keller, J.B. (1968). Active leisure time activities as related to age among males in a total population. *Journal of Gerontology,* **23,** 351-356.

Cunningham, D.A., Montoye, H.J., Metzner, H.L., & Keller, J.B. (1969). Physical activity at work and active leisure as related to occupation. *Medicine and Science in Sports,* **1,** 165-170.

Dalsky, G.P., Stocke, K.S., Ehsani, A.A., Slatopolsky, E., Lee, W.C., & Birge, S.J. (1988). Weight-bearing exercise training and lumbar bone mineral content in postmenopausal women. *Annals of Internal Medicine,* **108,** 824-828.

Dannenberg, A.L., Keller, J.B., Wilson, P.W.F., & Castelli, W.P. (1989). Leisure time physical activity in the Framingham Offspring Study: Description, seasonal variation and risk factor correlates. *American Journal of Epidemiology,* **129,** 76-88.

Dannenberg, A.L., & Wilson, P.W.E. (1989). Framingham Leisure Time Physical Activity Questionnaire. In T.M. Drury (Ed.), *Assessing physical fitness and physical activity in population-based surveys* (pp. 649-652; DHHS Pub. No. PHS 89-1253). Washington, DC: U.S. Government Printing Office.

DeBacker, G., Kornitzer, M., Sobolski, J., Dramaix, M., Degré, S., DeMarneffe, M., & Denolin, H. (1981). Physical activity and physical fitness levels of Belgian males aged 40-55 years. *Cardiology,* **67,** 110-128.

DiPietro, L., Caspersen, C.J., Ostfeld, A.M., & Nadel, E.R. (1993). A survey for assessing physical activity among older adults. *Medicine and Science in Sports and Exercise,* **25,** 628-642.

Dishman, R.K. (1988). Supervised and free-living physical activity: No differences in former athletes and non-athletes. *American Journal of Preventive Medicine,* **4,** 153-160.

Division of Occupational Analysis, U.S. Employment Service. (1949). *Dictionary of occupational titles* (Vol. 2). Washington, DC: U.S. Government Printing Office.

Durant, R.H., Linder, C.W., Harkness, J.W., & Gray, R.G. (1983). Relationship between physical activity and serum lipids and lipoproteins in black children and adolescents. *Journal of Adolescent Health Care,* **4,** 55-60.

Durant, R.H., Linder, C.W., & Mahoney, O.M. (1983). Relationship between habitual physical activity and serum lipoprotein levels in white male adolescents. *Journal of Adolescent Health Care,* **4,** 235-240.

Engström, L.-M. (1980). Physical activity of children and youth. *Acta Paediatrica Scandinavica,* **283**(Suppl.), 101-105.

Epstein, L., Miller, G.J., Stitt, F.W., & Morris, J.N. (1976). Vigorous exercise in leisure time, coronary risk factors, & resting electrocardiogram in middle aged male servants. *British Heart Journal* **38,** 403-409.

Folsom, A.R., Jacobs, D.R., Jr., Caspersen, C.J., Gomez-Marin, O., & Knudsen, J., (1986). Test-retest reliabilities of the Minnesota Leisure Time Physical Activity Questionnaire. *Journal of Chronic Disease,* **39,** 505-511.

Frank, C.W., Weinblatt, E., Shapiro, S., & Sager, R.V. (1966a). Myocardial infarction in men: Role of physical activity and smoking in incidence and mortality. *Journal of the American Medical Association,* **198,** 1241-1245.

Frank, C.W., Weinblatt, E., Shapiro, S., & Sager, R.V. (1966b). Physical inactivity as a lethal factor in myocardial infarction among men. *Circulation,* **34,** 1022-1033.

Frisch, R.E., Wyshak, G., Albright, N.L., Albright, T.E., Schiff, I., Jones, K.P., Witschi, J., Shiang, E., Koff, E., & Marguglio, M. (1985). Lower prevalence of breast cancer and cancers of the reproductive system among former college athletes compared to non-athletes. *British Journal of Cancer,* **52,** 885-891.

Frisch, R.E., Wyshak, G., Albright, T.E., Albright, N.L., & Schiff, I. (1986). Lower prevalence of diabetes in female former college athletes compared with non-athletes. *Diabetes,* **35,** 1101-1105.

Frisch, R.E., Wyshak, G., Witschi, J., Albright, N.L., Albright, T.E., & Schiff, I. (1987). Lower lifetime occurrence of breast cancer and cancers of the reproductive system among former college athletes. *International Journal of Fertility,* **32,** 217-225.

Garcia-Palmieri, M.R., Costas, R., Cruz-Vidal, M., Sorlie, P.D., & Havlik, R.J. (1982). Increased physical activity: A protective factor against heart attacks in Puerto Rico. *American Journal of Cardiology,* **50,** 749-755.

Gionet, N.J., & Godin, G. (1989). Self-reported exercise behavior of employees: A validity study. *Journal of Occupational Medicine,* **31,** 969-973.

Godin, G., & Shephard, R.J. (1984). Normative beliefs of school children concerning regular exercise. *Journal of School Health,* **54,** 443-445.

Godin, G., & Shephard, R.J. (1985). A simple method to assess exercise behavior in the community. *Canadian Journal of Applied Sport Science,* **10,** 141-146.

Gordon, D.J. Leon, A.S., Ekelund, L.-G., Sopko, G., Probstfield, J.L., Rubenstein, C., & Sheffield, L.T. (1987). Smoking, physical activity, and other predictors of endurance and heart rate response to exercise in asymptotic hypercholesterolemic men. *American Journal of Epidemiology,* **125,** 587-600.

Gordon, D.J., Witztum, J.L., Hunninghake, D., Gates, S., & Glueck, C.J. (1983). Habitual physical activity and high-density lipoprotein cholesterol in men with primary hypercholesterolemia. *Circulation,* **67,** 512-520.

Gretebeck, R.J., Montoye, H.J. & Porter, W. (1993). Validation of a portable accelerometer for estimating energy expenditure using doubly labeled water. Manuscript submitted for publication.

Grimby, G., Wilhelmsen, L., Björntorp, P.,Saltin, B., & Tibblin, G. (1971). Habitual physical activity: Aerobic power and blood lipids. In B. Pernow & B. Saltin (Eds.), *Muscle metabolism during exercise* (pp. 469-481). New York: Plenum Press.

Gross, L.D., Sallis, J.F., Buono, M.J., Roby, J.J., & Nelson, J.A. (1990). Reliability of interviewers using the seven-day physical activity recall. *Research Quarterly for Exercise and Sport,* **61,** 321-325.

Gyntelberg, F. & Ohlsen, K. (1973). Physical fitness and serum cholesterol in Copenhagen males aged 40-59. *Scandinavian Journal of Clinical Laboratory Investigation,* **32,** 211-216.

Halverson, C.F., & Waldrop, M.F. (1973). The relation of mechanically recorded activity level to varieties of preschool play behavior. *Child Development,* **44,** 678-681.

Haskell, W.L., Taylor, H.L., Wood, P.D., Schrott, H., & Heiss, G. (1980) Strenuous physical activity, treadmill exercise test performance and plasma high-density lipoprotein cholesterol. *Circulation,* **62** (Suppl. IV), 53-61.

Health and Welfare Canada and Statistics Canada. (1981). *The health of Canadians: Report of the Canada Health Survey* (pp. 71-76). Ottawa, ON: Minister of Supply and Services.

Hedley, O.F. (1939). Analysis of 5,116 deaths reported as due to acute coronary occlusion in Philadelphia, 1933-1937. *U.S. Weekly Public Health Reports,* **54,** 972-1012.

Hennekens, C.H., Rosner, B., Jesse, M.J., Drolette, M.E., & Speizer, F.E. (1977). A retrospective study of physical activity and coronary deaths. *International Journal of Epidemiology,* **6,** 243-246.

Hickey, N., Mulcahy, R., Bourke, G.J., Graham, I., & Wilson-Davis, K. (1975). Study of coronary risk factors related to physical activity in 15,171 men. *British Medical Journal,* **3,** 507-509.

Holme, I., Helgeland, A., Hjermann, I., Leren, P., & Lund-Larsen, P.G. (1981). Physical activity at work and at leisure in relation to coronary risk factors and social class. *Acta Medica Scandinavica,* **209,** 277-283.

Hopkins, W.G., Wilson, N.C., & Russell, D.G. (1991). Validation of the physical activity instrument for the life in New Zealand National Survey. *American Journal of Epidemiology,* **133,** 73-82.

Jacobs, D.R., Jr. (1988). *A revision of the Tecumseh Leisure Time Physical Activity Interview.* Unpublished manuscript.

Jacobs, D.R., Jr., Ainsworth, B.E. Hartman, T.J., & Leon, A.S. (1993). A simultaneous evaluation of ten commonly used physical activity questionnaires. *Medicine and Science in Sports and Exercise,* **25,** 81-91.

Jacobs, D.R., Jr., & Montoye, H.J. (1988). *A revision of the Tecumseh Occupational Physical Activity Interview.* Unpublished manuscript.

Janz, K.F., Phillips, A., & Mahoney, L.T. (1992). Self-selected physical activity profiles in children and adolescents. *Physical Educator,* **49,** 81-87.

Kannel, W.B., & Sorlie, P. (1979). Some health benefits of physical activity. The Framingham study. *Archives of Internal Medicine,* **139,** 857-861.

Kaplan, G.A., Seeman, T.E., Cohen, R.D., Knudsen, L.P., & Guralnik, J. (1987). Mortality among the elderly in the Alameda County Study: Behavioral and demographic risk factors. *American Journal of Public Health,* **77,** 307-312.

Kemper, H.C.G. (1992). Physical development and childhood physical activity. In N.G. Norgan (Ed.), *Physical activity and health* (pp. 84-101). Cambridge: Cambridge University Press.

Kemper, H.C.G., Suel, J., Verschuur, R., & Storm-van Essen, L. (1990). Tracking of health and risk indicators of cardiovascular diseases from teenager to adult: Amsterdam Growth and Health Study. *Preventive Medicine,* **19,** 642-655.

Kemper, H.C.G., & Verschuur, R. (1974). Relationship between biological age, habitual physical activity and morphological, physiological characteristics of 12 and 13-year-old boys. *Acta Paediatrica Belgica,* **28** (Suppl.), 191-203.

Klesges, R.C., Eck, L.H., Mellon, M.W., Fulliton, W., Somes, G.W., & Hanson, C.L. (1990). The accuracy of self-reports of physical activity. *Medicine and Science in Sports and Exercise,* **22,** 690-697.

Kohl, H.W., Blair, S.N., Paffenbarger, R.S., Jr., Macera, C.A., & Kronenfeld, J.J. (1988). A mail survey of physical habits as related to measured physical fitness. *American Journal of Epidemiology,* **127,** 1228-1239.

Kohl, H.W., Harris, M.J., & Blair, S.N. (1993). Assessment of historical physical activity behavior in women and men using a mail survey [abstract]. *Medicine and Science in Sports and Exercise,* **25**(Suppl.), S27.

Kriska, A.M., Black-Sandler, R. Cauley, J.A., LaPorte, R.E., Hom, D.L., and Pambianco, G. (1988). *American Journal of Epidemiology,* **127,** 1053-1063.

Kriska, A.M., Knowler, W.C., LaPorte, R.E., Drash, A.L., Wing, R.R., Blair, S.N., Bennett, P.H., & Kuller, L.H. (1990). Development of a questionnaire to examine relationship of physical activity and diabetes in Pima Indians. *Diabetes Care,* **13,** 401-411.

Kriska, A.M., LaPorte, R.E., & Knowler, W.C. (1992). The association of physical activity, obesity, fat distribution and glucose tolerance in Pima Indians [abstract]. *Medicine and Science in Sports and Exercise,* **24,** S61.

Laakso, L., & Telama, R. (1981). Sport participation of Finnish youth as a function of age and schooling. *Sportwissenschaft,* **11,** 28-45.

Lamb, K.L., & Brodie, D.A. (1991). Leisure-time physical activity as an estimate of physical fitness: A validation study. *Journal of Clinical Epidemiology,* **44,** 41-52.

Lapidus, L., & Bengtsson, C. (1986). Socioeconomic factors and physical activity in relation to cardiovascular disease and death. *British Heart Journal,* **55,** 295-301.

LaPorte, R.E., Cauley, J.A., Kinsey, C.M., Corbett, W., Robertson, R., Black-Sandler, R., Kuller, L.H., & Falkel, J. (1982). The epidemiology of physical activity in children, college students, middle-aged men, menopausal females and monkeys. *Journal of Chronic Diseases,* **35,** 787-795.

Lee, I.-M., Hsieh, C.-C., & Paffenbarger, R.S., Jr. (1993). Vigorous physical activity, non-vigorous physical activity, and risk of mortality in men [abstract]. *Medicine and Science in Sports and Exercise,* **25,** S167.

Lee, I.-M., Paffenbarger, R.J., Jr., & Hsieh, C.-C. (1992). Time trends in physical activity among college alumni, 1962-1988. *American Journal of Epidemiology,* **135,** 915-925.

Lehmann, G. (1953). *Praktische Arbeitsphysiologie* [Practical Work Physiology]. Stuttgart: Thieme.

Leon, A.S., Connett, J., Jacobs, D.R., Jr., & Rauramaa, R. (1987). Leisure time physical activity levels and risk of coronary heart disease and death. *Journal of the American Medical Association,* **258,** 2388-2395.

Leon, A.S., Connett, J., & the MRFIT Research Group. (1991). Physical activity and 10.5 year mortality in the multiple risk factor intervention trial (MRFIT). *International Journal of Epidemiology,* **20,** 690-697.

Leon, A.S., Jacobs, D.R., Jr., DeBacker, G., & Taylor, H.L. (1981). Relationship of physical characteristics and life habits to treadmill exercise capacity. *American Journal of Epidemiology,* **113,** 653-660.

Linder, C.W., Durant, R.H., & Mahoney, O.M. (1983). The effect of physical conditioning on serum lipids and lipoproteins in white male adolescents. *Medicine and Science in Sports and Exercise,* **15,** 232-236.

Ludvigsson, J. (1980). Physical exercise in relation to degree of metabolic control in juvenile diabetics. *Acta Paediatrica Scandinavica,* **283**(Suppl.), 45-49.

Magnus, K., Matroos, A., & Strackee, J. (1979). Walking, cycling or gardening, with or without seasonal interruption, in relation to acute coronary events. *American Journal of Epidemiology,* **110,** 724-733.

Mahoney, M., & Freedson, P. (1990). Assessment of physical activity from Caltrac and Baecke questionnaire techniques [abstract]. *Medicine and Science in Sports and Exercise,* **22,** S80.

Marr, J.W., Gregory, J., Meade, T.W., Alderson, M.R., & Morris, J.N. (1970). Diet, leisure activity and skinfold measurements in sedentary men [abstract]. *Proceedings of the Nutrition Society,* **29,** 17A.

Marti, B., & Vaartiainen, E. (1989) Relation between leisure time exercise and cardiovascular risk factors among 15-year-olds in eastern Finland. *Journal of Epidemiology and Community Health,* **43,** 228-233.

Miller, D.J., Freedson, P.S., & Kline, G.M. (1994). Comparison of activity levels using Caltrac® accelerometer and five questionnaires. *Medicine and Science in Sports and Exercise,* **26,** 376-382.

Moisan, J., Meyer, F., & Gingras, S. (1991). Leisure physical activity and age at menarche. *Medicine and Science in Sports and Exercise,* **23,** 1170-1175.

Montoye, H.J. (1971). Estimation of habitual physical activity by questionnaire and interview. *American Journal of Clinical Nutrition,* **24,** 1113-1118.

Montoye, H.J. (1975). *Physical activity and health: An epidemiologic study of an entire community.* Englewood Cliffs, NJ: Prentice-Hall.

Montoye, H.J. (1985). Risk indicators for cardiovascular disease in relation to physical activity in youth. In R.A. Binkhorst, H.C.G. Kemper, & W.H.M. Saris (Eds.), *Children and exercise XI* (pp. 3-25). Champaign, IL: Human Kinetics.

Montoye, H.J. (1986). Physical activity, physical fitness, and heart disease risk factors in children. *The Academy Papers,* **19,** 127-152.

Montoye, H.J., Block, W.D., Metzner, H.L., & Keller, J.B. (1976). Habitual physical activity and serum lipids: Males, age 16-64 in a total community. *Journal of Chronic Disease,* **29,** 697-709.

Montoye, H.J., Block, W.D., Metzner, H.L., & Keller, J.B. (1977). Habitual physical activity and glucose tolerance: Males age 16-64 in a total community. *Diabetes,* **26,** 172-176.

Montoye, H.J., Metzner, H.L., Keller, J.B., Johnson, B.C., & Epstein, H.E. (1972). Habitual physical activity and blood pressure. *Medicine and Science in Sports,* **4,** 175-181.

Montoye, H.J., Mikkelsen, W.H., Metzner, H.L., & Keller, J.B. (1976). Physical activity, fatness, and serum uric acid. *Journal of Sports Medicine and Physical Fitness,* **16,** 253-260.

Mor, V., Murphy, J., Masterson-Allen, S., Wiley, C., Razmpour, A., Jackson, M.E., Greer, D., & Katz, S. (1989). Risk of functional decline among well elders. *Journal of Clinical Epidemiology,* **42,** 895-904.

Morris, J.N., Chave, S.P.W., Adam, C., & Sirey, C. (1973). Vigorous exercise in leisure time and incidence of coronary heart disease. *Lancet,* **1,** 333-339.

Morris, J.N., Everitt, M.G., Pollard, R., & Chave, S.P.W. (1980). Vigorous exercise in leisure time: Protection against coronary heart disease. *Lancet,* **2,** 1207-1210.

Morris, J.N., Heady, J.A., Raffle, P.A.B., Roberts, C.G., & Parks, J.W. (1953). Coronary heart disease and physical activity of work. *Lancet,* **265,** 1053-1057, 1111-1120.

Mundal, R., Erikssen, J., & Rodahl, K. (1987). Assessment of physical activity by questionnaire and personal interview with particular reference to fitness and coronary mortality. *European Journal of Applied Physiology,* **56,** 245-252.

Murphy, J.K. Alpert, B.S., Christman, J.V., & Willey, E.S. (1988). Physical fitness in children: A survey method based on parental report. *American Journal of Public Health,* **78,** 708-710.

National Center for Health Statistics. (1978). Exercise and participation in sports among persons 20 years of age and over: United States, 1975. *Vital and Health Statistics, Advance Data,* Number 19.

National Center for Health Statistics. (1980). Health practices among adults: United States, 1977. *Vital and Health Statistics, Advance Data,* **64.**

National Center for Health Statistics. (1981a). Highlights from Wave I of the National Survey of Personal Health Practices and Consequences: United States, 1979. *Vital and Health Statistics,* Series 15, No. 1.

National Center for Health Statistics. (1981b). Plan and operation of the second National Health and Nutrition Examination Survey, 1976-1980. *Vital and Health Statistics,* Series 1, No. 15.

National Office of Vital Statistics, U.S. Department of Health, Education and Welfare. (1957). *A physical activity scale assigned to the detailed occupation code for males.* Washington, DC: U.S. Government Printing Office.

Noland, M., Danner, F., Dewalt, K., McFadden, M., & Kotchen, J.M. (1990). The measurement of physical activity in young children. *Research Quarterly for Exercise and Sports,* **61,** 146-153.

Owen, N., Sedgwick, H.W., & Davies, M. (1988). Validity of a simplified measure of participation in vigorous physical activity. *Medical Journal of Australia,* **148,** 600.

Paffenbarger, R.S. (1985). Physical activity as a defense against coronary heart disease. In W.E. Connor & J.D. Bristow (Eds.), *Coronary heart disease: Prevention, complications, and treatments* (pp. 135-155). Philadelphia: J.B. Lippincott.

Paffenbarger, R.S., Jr., Blair, S.N., Lee, I.-M., & Hyde, R.T. (1993). Measurement of physical activity to assess health effects in free-living populations. *Medicine and Science in Sports and Exercise,* **25,** 60-70.

Paffenbarger, R.S., Jr., Hyde, R.T., & Wing, A.L. (1987). Physical activity and incidence of cancer in diverse populations: A preliminary report. *American Journal of Clinical Nutrition,* **45,** 312-317.

Paffenbarger, R.S., Jr., Hyde, R.T., Wing, A.L., & Hsieh, C.C. (1986). Physical activity, all-cause mortality, and longevity of college alumni. *New England Journal of Medicine,* **314,** 605-613.

Paffenbarger, R.S., Jr., Hyde, R.T., Wing, A.L., & Steinmetz, C.H. (1984). A natural history of athleticism and cardiovascular health. *Journal of the American Medical Association,* **252,** 491-495.

Paffenbarger, R.S., Jr., Hyde, R., Wing, A., Jung, D., & Kampert, J. (1991). Influences of changes in physical activity and other characteristics on all-cause mortality [abstract]. *Medicine and Science in Sports and Exercise,* **23,** S82.

Paffenbarger, R.S., Jr., Laughlin, M.E., Gima, A.S., & Black, R.A. (1970). Work activity of longshoremen as related to death from coronary heart disease and stroke. *New England Journal of Medicine,* **282,** 1109-1114.

Paffenbarger, R.S., Jr., Wing, A.L., & Hyde, R.T. (1978). Physical activity as an index of heart attack risk in college alumni. *American Journal of Epidemiology,* **108,** 161-175.

Paffenbarger, R.S., Jr., Wing, A.L., Hyde, R.T., & Jung, D.L. (1983). Physical activity and incidence of hypertension in college alumni. *American Journal of Epidemiology,* **117,** 245-257.

Pate, R.R., Dowda, M., & Ross, J.G. (1990). Associations between physical activity and physical fitness in American children. *American Journal of Diseases of Children,* **144,** 1123-1129.

Pearl, R. (1924). *Studies of human biology.* Baltimore: Williams and Wilkins.

Powell, K.E., Thompson, P.D., Caspersen, C.J., & Kendrick, J.S. (1987). Physical activity and the incidence of coronary heart disease. *Annual Review of Public Health,* **8,** 253-387.

President's Council on Physical Fitness and Sports. (1974). National adult physical fitness survey. *Physical Fitness Research Digest,* **4,** 1-27.

Purdue Farm Cardiac Project. (1961). Energy requirements for physical work. *Agriculture experiment station mimeographed research progress report no. 30.* Lafayette, IN: Purdue University Press.

Rauh, M.J.D., Hovell, M.F., Hofstetter, C.R., Sallis, J.F., & Gleghorn, A. (1992). Reliability and validity of self-reported physical activity in Latinos. *International Journal of Epidemiology,* **21,** 966-971.

Reaven, P.D., McPhillips, J.B., Barrett-Connor, E.L., & Criqui, M.H. (1990). Leisure time exercise and lipid and lipoprotein levels in an older population. *Journal of the American Geriatric Society,* **38,** 847-854.

Reiff, G.G., Montoye, H.J., Remington, R.D., Napier, J.A., Metzner, H.L., & Epstein, F.H. (1967). Assessment of physical activity by questionnaire and interview. *Journal of Sports Medicine and Physical Fitness,* **7,** 135-142.

Riddoch, C. (1990). *Northern Ireland health and fitness survey.* Belfast, Northern Ireland: The Queens University of Belfast.

Roby, J.J., Sallis, J.F., Kolodoy, B., Condon, A., & Goggin, K. (1992). Developing self-administered self-reports of children's physical activity [abstract]. *Medicine and Science in Sports and Exercise,* **24,** S69.

Ross, C.E., & Hayes, D. (1988). Exercise and psychologic well-being in the community. *American Journal of Epidemiology,* **127,** 762-771.

Ross, I.G. (1989). Evaluating fitness and activity assessments from the National Children and Youth Fitness Studies I and II. In T.F. Drury (Ed.), *Assessing physical fitness and physical activity in population-based surveys* (pp. 229-259; DHH Pub. No. PHS 89-1253). Washington, DC: U.S. Government Printing Office.

Sallis, J.F. (1991). Self-report measures of children's physical activity. *Journal of School Health,* **61,** 215-219.

Sallis, J.F., Buono, M.J., & Freedson, P.S. (1991). Bias in estimating caloric expenditure from physical activity in children. *Sports Medicine,* **11,** 203-209.

Sallis, J.F., Buono, M.J., Roby, J.J., Carlson, D., & Nelson, J.A. (1990). The Caltrac accelerometer as a physical activity monitor for school-age children. *Medicine and Science in Sports and Exercise,* **22,** 698-703.

Sallis, J.F., Buono, M.J., Roby, J.J., Micale, F.G., & Nelson, J.A. (1993). Seven-day recall and other physical activity self-reports in children and adolescents. *Medicine and Science in Sports and Exercise,* **25,** 99-108.

Sallis, J.F., Haskell, W.L., Wood, P.D., Fortmann, S.P., Rogers, T., Blair, S.N., & Paffenbarger, R.S., Jr. (1985). Physical activity assessment methodology in the five-city project. *American Journal of Epidemiology,* **121,** 91-106.

Sallis, J.F., Patterson, T.L., Buono, M.J., Atkins, C.J., & Nader, P.R. (1988). Aggregation of physical activity habits in Mexican-Americans and Anglo families. *Journal of Behavioral Medicine,* **11,** 31-41.

Sallis, J.F., Patterson, T.C., Buono, M.J., & Nader, P.R. (1988). Relation of cardiovascular fitness and physical activity to cardiovascular disease risk factors in children and adults. *American Journal of Epidemiology,* **127,** 933-941.

Salonen, J.T., Puska, P., & Tuomilehto, J. (1982). Physical activity and risk of myocardial infarction, cerebral stroke and death. *American Journal of Epidemiology,* **115,** 526-537.

Saltin, B., & Grimby, G. (1968). Physiological analysis of middle-aged and old former athletes. Comparison with still active athletes of the same ages. *Circulation,* **38,** 1104-1115.

Saris, W.H.M., & Binkhorst, R.A. (1977). The use of pedometer and actometer in studying daily physical activity in man. Part II: Validity of pedometer and actometer measuring daily physical activity. *European Journal of Applied Physiology,* **37,** 229-235.

Saris, W.H.M., Binkhorst, R.A., Cramwinckel, A.B., Van der Veen-Hezemans, A.M., & Van Waesberghe, F. (1979). Evaluation of somatic effects of a health education program for school children. *Bibliotheca Nutritio et Dieta,* **27,** 77-84.

Saris, W.H.M., Doesburg, W.H., Lemmens, W.A.J.G., & Reingis, A. (1974). *Habitual physical activity in children: Results of a questionnaire and movement counters,* (pp. III79-III92). Niymegen, The Netherlands: Report of the Health Education Project (GVO).

Saris, W.H.M., Westerterp, K.R., Kempen, K., Bloemberg, B.P.M., Caspersen, C., & Kromhout, D. (1993). Validation of the Zutphen physical activity questionnaire among elderly men by the doubly labeled water technique. Unpublished manuscript.

Schechtman, K.B., Barzilai, B., Rost, K., & Fisher, E.B., Jr. (1991). Measuring physical activity with a single question. *American Journal of Public Health,* **81,** 771-773.

Schmücker, B., Rigauer, B., Hinrichs, W., & Trawinski, J. (1984). Motor abilities and habitual physical activity in children. In J. Ilmarinen & I. Välimäki (Eds.), *Children and sport* (pp. 46-52). Berlin: Springer-Verlag.

Shapiro, S., Weinblatt, E., Frank, C.W., & Sager, R.V. (1965). The H.I.P. study of incidence and prognosis of coronary heart disease: Preliminary findings on incidence of myocardial infarction and angina. *Journal of Chronic Disease,* **18,** 527-558.

Shapiro, S., Weinblatt, E., Frank, C.W., Sager, R.V., & Densen, P.M. (1963). The H.I.P. study of incidence and diagnosis of coronary heart disease: Methodology. *Journal of Chronic Disease,* **16,** 1281-1292.

Shaw, S.M. (1985). Gender and leisure: Inequality in the distribution of leisure time. *Journal of Leisure Research,* **17,** 266-282.

Siconolfi, S.F., Lasater, T.M., Snow, R.C.K., & Carleton, R.A. (1985). Self-reported physical activity compared with maximal oxygen uptake. *American Journal of Epidemiology,* **122,** 101-105.

Simons-Morton, B.G., O'Hara, N.M., Parcel, G.S., Huang, I.W., Baranowski, T., & Wilson, B. (1990). Children's frequency of participation in moderate to vigorous physical activities. *Research Quarterly for Exercise and Sports,* **61,** 307-314.

Skinner, J.S., Benson, H., McDonough, J.R., & Hames, C.G. (1966). Social status, physical activity, and coronary proneness. *Journal of Chronic Diseases,* **19,** 773-783.

Sobolski, J., Kornitzer, M., DeBacker, G., Dramaix, M., Abramowicz, M., Degré, S., & Denolin, A., (1987). Protection against ischemic heart disease in the Belgian physical fitness study: Fitness rather than physical activity? *American Journal of Epidemiology,* **125,** 601-610.

Sorock, G.S., Bush, T.L., Golden, A.L., Fried, L.P., Breuer, B., & Hale, W.E. (1988). Physical activity and fracture risk in a free-living elderly cohort. *Journal of Gerontology,* **43,** M134-139.

Stephens, T. (1983). *Fitness and lifestyle in Canada.* Ottawa, ON: Canadian Fitness and Lifestyle Research Institute.

Stephens, T., & Craig, C.L. (1989). Fitness and activity measurement in the 1981 Canada Fitness Survey. In T.F. Drury (Ed.), *Assessing physical fitness and physical activity in population-based surveys* (pp. 401-432; DHHS Pub. No. PHS 89-1253). Washington, DC: U.S. Government Printing Office.

Stephens, T., Craig, C.L., & Ferris, B.F. (1986). Adult physical activity in Canada: Findings from the Canadian Fitness Survey I. *Canadian Journal of Public Health,* **77,** 285-290.

Stillman, R.J., Lohman, T.G., Slaughter, M.H., & Massey, B.H. (1986). Physical activity and bone mineral content in women aged 30-85 years. *Medicine and Science in Sports and Exercise,* **18,** 576-580.

Strazzullo, P., Cappuccio, F.P., Trevisan, M., DeLeo, A., Krogh, V., Giorgione, N., & Mancini, M. (1988). Leisure time physical activity and blood pressure in school children. *American Journal of Epidemiology,* **127,** 726-733.

Sunnegårdh, J., & Bratteby, L.-E. (1987). Maximal oxygen uptake, anthropometry and physical activity in randomly selected sample of 8 and 13-year-old children in Sweden. *European Journal of Applied Physiology,* **56,** 266-272.

Suter, E., & Hawes, M.R. (1993). Relationship of physical activity, body fat, diet, and blood lipid profile in youths 10-15 years. *Medicine and Science in Sports and Exercise,* **25,** 748-754.

Taylor, C.B., Coffey, T., Berra, K., Iaffaldano, R., Casey, K., & Haskell, W.L. (1984). Seven-day activity and self-report compared to a direct measure of physical activity. *American Journal of Epidemiology,* **120,** 818-824.

Taylor, H.L., Jacobs, D.R., Schucker, B., Knudsen, J., Leon, A.S., & DeBacker, G. (1978). A questionnaire for the assessment of leisure time physical activities. *Journal of Chronic Disease,* **31,** 741-755.

Taylor, H.L., Klepetar, E., Keys, A., Parlin, W., Blackburn, H., & Puchner, T. (1962). Death rates among physically active and sedentary employees of the railroad industry. *American Journal of Public Health,* **52,** 1697-1707.

Tell, G.S., & Vellar, O.D. (1988). Physical fitness, physical activity, and cardiovascular disease risk factors in adolescents: The Oslo youth study. *Preventive Medicine,* **17,** 12-24.

The Gallup poll. (1984). *Six of ten adults exercise regularly.* The Los Angeles Times Syndicate, May.

The Miller Lite Report on American Attitudes Toward Sports. (1983). Milwaukee, WI: Miller Brewing.

The Perrier Survey: Fitness in America. (1979). New York: Perrier—Great Waters of France.

Uitenbroek, D.G. (1993). Seasonal variation in leisure time physical activity. *Medicine and Science in Sports and Exercise,* **25,** 755-760.

U.S. Employment Service. (1955). *Estimates of worker trait requirements for 4,000 jobs.* Washington, DC: U.S. Government Printing Office.

Verschuur, R., & Kemper, H.C.G. (1985a). Habitual physical activity. *Medicine and Sports Science,* **20,** 56-65.

Verschuur, R., & Kemper, H.C.G. (1985b). The pattern of daily physical activity. *Medicine and Sports Science,* **20,** 169-186.

Verschuur, R., & Kemper, H.C.G. (1987). Longitudinal changes in daily physical activity in girls and boys between age 12 and 18. In R. Verschuur (Ed.), *Daily physical activity and health* (pp. 61-94). Haarlem, The Netherlands: B.V. Uitgeverij de Vrieseboch.

Verschuur, R., Kemper, H.C.G., & Storm-van Essen, L. (1987). Longitudinal changes in habitual physical activity, and in physical fitness, body size and body composition of boys at age 12/13 and 16/17. In R. Verschuur (Ed.), *Daily physical activity and health* (pp. 21-44). Haarlem, The Netherlands: B.V. Uitgeverij de Vrieseboch.

Verschuur, R., Ritmeester, J.W., Kemper, H.C.G., & Storm-van Essen, L. (1987). Daily physical activity, sports participation and cardiovascular disease risk indicators in teenagers. In R. Verschuur (Ed.), *Daily physical activity and health* (pp. 158-172). Haarlem, The Netherlands: B.V. Uitgeverig de Vrieseboch.

Viikari, J., Välimäki, I., Telama, R., Siren-Tiusanen, H., Åkerblom, H.K., Dahl, M., Lähde, P.-L., Pesonen, E., Pietikäinen, M., Suoninen, P., & Uhari, M. (1984). Atherosclerosis precursors in Finnish children: Physical activity and plasma lipids in 3 and 12-year old children. In J. Ilmarinen & I. Välimäki (Eds.), *Children and sport* (pp. 231-240). Berlin: Springer-Verlag.

Vogel, J.A. (1989). Fitness and activity assessments among U.S. Army populations: Implications for NCHS general population surveys. In T.F. Drury (Ed.), *Assessing physical fitness and physical activity in population-based surveys* (pp. 377-399; DHHS Pub. No. PHS 89-1253). Washington, DC: U.S. Government Printing Office.

Voorrips, L.E., Ravelli, A.C.J., Dongelmans, P.C.A., Deurenberg, P., & Van Staveren, W.A. (1991). A physical activity questionnaire for the elderly. *Medicine and Science in Sports and Exercise,* **23,** 974-979.

Voulle, P., Telama, R., & Laakso, L. (1986). Physical activity in the life style of Finnish adults. *Scandinavian Journal of Sports Science,* **8,** 105-115.

Wallace, J.P., McKenzie, T.L., & Nader, P.R. (1985). Observed versus recalled exercise behavior: A validation of a seven day exercise recall for boys 11 to 13 years old. *Research Quarterly for Exercise and Sports,* **56,** 161-165.

Wanne, O., Viikari, J., Telema, R., Åkerblom, H.K., Pesonen, E., Uhari, M., Dahl, M., Suoninen, P., & Välimäki, I. (1983). Physical activity and serum lipids in 8-year-old Finnish school boys. *Scandinavian Journal of Sports Science,* **5,** 10-14.

Washburn, R.A., Goldfield, S.R.W., Smith, K.W., & McKinlay, J.B. (1990). The validity of self-reported exercise-induced sweating as a measure of physical activity. *American Journal of Epidemiology,* **132,** 107-113.

Washburn, R.A., Jette, A.M., & Janney, C.A. (1990). Using age-neutral physical activity questionnaires in research with the elderly. *Journal of Aging and Health,* **2,** 341-356.

Washburn, R.A., & Montoye, H.J. (1986) The assessment of physical activity by questionnaire. *American Journal of Epidemiology,* **123,** 563-575.

Washburn, R.A., Smith, K.W., Goldfield, S.R.W., and McKinlay, J.B. (1991). Reliability and physiologic correlates of the Harvard Alumni Activity Survey in a general population. *Journal of Clinical Epidemiology,* **44,** 1319-1326.

Washburn, R.A., Smith, K.W., Jette, A.M., & Janney, C.A. (1993). The Physical Activity Scale for the Elderly (PASE): Development and evaluation. *Journal of Clinical Epidemiology,* **46,** 153-162.

Weiss, T.W., Slater, C.H., Green, L.W., Kennedy, V.C., Albright, D.L., & Wun, C.-C. (1990). The validity of single-item, self-assessment questions of measures of adult physical activity. *Journal of Clinical Epidemiology,* **43,** 1123-1129.

Wessel, J.A., Montoye, H.J., & Mitchell, M.A. (1965). Physical activity assessment by recall record. *American Journal of Public Health,* **55,** 1430-1436.

Wilbur, J., Miller, A., Dan, A.J., & Holm, K. (1989). Measuring physical activity in midlife women. *Public Health Nursing,* **6,** 120-128.

Wiley, J.A., & Camacho, T.C. (1980). Lifestyle and future health: Evidence from the Alameda County study. *Preventive Medicine,* **9,** 1-21.

Wilhelmsen, L. (1969). The myocardial infarction clinic in Göteborg: Organization and preliminary results. *Pehr Dubb Journalen,* **3,** 43-54.

Wilhelmsen, L., & Tibblin, G. (1970). Physical inactivity and risk of myocardial infarction: The men born in 1913 study. In O.A. Larson & R.O. Malmborg (Eds.), *Coronary heart disease and physical fitness* (pp. 251-255). Copenhagen: Munksgaard.

Williams, E., Klesges, R.C., Hanson, C.L., & Eck, L.H. (1989). A prospective study of the reliability and convergent validity of three physical activity measures in a field research trial. *Journal of Clinical Epidemiology,* **42,** 1161-1170.

Wingard, D.L., Berkman, L.F., & Brand, R.J. (1982). A multivariate analysis of health-related practices: A nine-year mortality follow up of the Alameda County study. *American Journal of Epidemiology,* **116,** 765-775.

Wyshak, G., Frisch, R.E., Albright, N.L., Albright, T.E., & Schiff, I. (1986). Lower prevalence of benign diseases of the breast and benign tumors of the reproductive system among former college athletes compared to nonathletes. *British Journal of Cancer,* **54,** 841-845.

Wyshak, G., Frisch, R.E., Albright, T.E., Albright, N.L. & Schiff, I. (1987). Bone fractures among former college athletes compared with nonathletes in the menopausal and postmenopausal years. *Obstetrics and Gynecology,* **69,** 121-126.

Yano, K., Reed, D.M., & McGee, D.L. (19484). Ten-year incidence of coronary heart disease in the Honolulu Heart Program: Relationship to biological and life-style characteristics. *American Journal of Epidemiology,* **119,** 653-666.

Yasin, S. (1967). Measuring habitual leisure-time physical activity by recall record questionnaire. In M.J. Karvonen & A.J. Barry (Eds.), *Physical activity and the heart* (pp. 372-373). Springfield, IL: Charles C Thomas.

Yasin, S., Alderson, M.R., Marr, J.W., Pattison, D.C., & Morris, J.N. (1967). Assessment of habitual physical activity apart from occupation.*British Journal of Preventive and Social Medicine,* **21,** 163-169.

Yeager, K.K. Macera, C.A., & Merritt, R.K. (1991). Sedentary women: Is it an issue of socioeconomic status? [abstract]. *Medicine and Science in Sports and Exercise,* **23,** S105.

York, E., Mitchell, R.E., & Graybiel, A. (1986). Cardiovascular epidemiology, exercise, and health: 40-year follow up of the U.S. Navy's 1,000 aviators. *Aviation, Space and Environmental Medicine,* **57,** 597-599.

Chapter 7

MOVEMENT ASSESSMENT DEVICES

The *pedometer* is a device to count steps taken or to estimate distance walked. Scientists may use very sophisticated pedometers, but even many average consumers are familiar with simpler versions to measure the extent of their activity. In this chapter we'll first discuss the pedometer and other instruments for counting steps, then move on to look at other kinds of movement counters, including portable accelerometers, which use transducers to measure the acceleration of the body in one or more directions. Although the pedometer depends on acceleration and deceleration, it is not an accelerometer in the true sense.

Pedometers

According to the *World Almanac of Presidential Facts,* Thomas Jefferson, the third president of the United States, invented the pedometer. However, the first pedometer was actually probably designed by Leonardo da Vinci about 500 years ago (Gibbs-Smith, 1978). A lever arm was likely meant to be attached to the thigh, and when the thigh moved back and forth with each step, the gears were rotated and the steps counted or distance estimated. In the museum of the university at Leiden in The Netherlands, one can see a pedometer used in the 17th century to count steps, apparently to measure plots of land. Another early pedometer, shown in Figure 7.1, was developed more than 100 years ago by the French physiologist Marey (Amar, 1920). A pneumatic sensor was attached to the sole of each shoe, and tubes connected the pads to a recording device held in the subject's hand. There is no evidence, to our knowledge, that this impractical instrument was ever used.

Modern Pedometers—Principles of Operation

The familiar modern pedometers are clipped to a belt or worn on the ankle. Used as early as 1926 (Lauter), they are designed to count steps in walking and possibly running; they should not be expected to measure other kinds of activities or total energy expenditure. Pedometers do not operate on the vertical pendulum principle, as is commonly believed. Rather, steps are counted in response to vertical acceleration of the body, which causes a lever arm to move vertically and a ratchet to rotate (Figure 7.2). This principle is employed in pedometers manufactured in the United States, Germany, Russia (Kemper & Verschuur, 1974) and Japan (Gayle, Montoye, & Philpot, 1977).

Several kinds of pedometers are illustrated in Figure 7.3. Some adjust for stride length to estimate distances walked. However, even with this adjustment, greater error is to be expected in estimating distance walked versus simply counting steps. A more sophisticated pedometer has recently been developed and is marketed by Panasonic of Japan. It is a battery-operated unit weighing 26 g (including battery) that clips onto a belt.

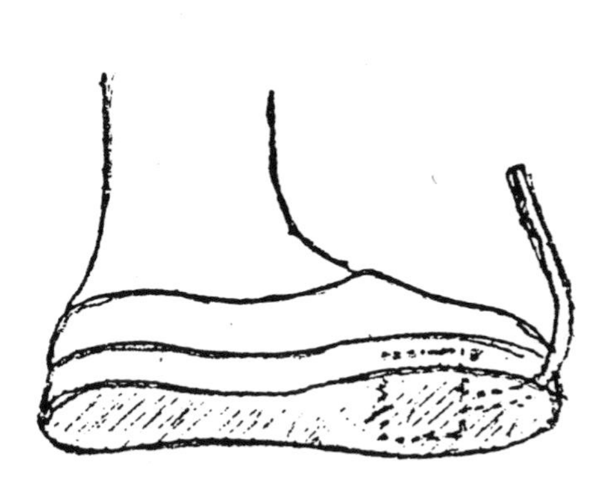

Figure 7.1 Marey's pedometer. From Amar, J. (1920). *The Human Motor*, p. 282, New York: E.P. Dutton.

It has not only a stride length adjustment but also a sensitivity adjustment. The number of steps may be recorded, and if one's weight is programmed into the unit, the readout can be in kilocalories. The unit sells for about two or three times the cost of ordinary pedometers. At this writing, neither its accuracy nor reliability has been reported.

Reliability and Validity

From data of Benedict and Murschhauser (1915) and Cotes and Meade (1960), Bassey, Dallosso, Fentem, Irving, and Patrick (1987) calculated that the maximum vertical acceleration of the hip in walking varies from about 0.5 $m \cdot s^{-2}$ to about 8 $m \cdot s^{-2}$. However, they have shown that the response of one brand of pedometer (Yamasa) quickly falls off to zero for accelerations below 2.5 $m \cdot s^{-2}$ and that there is a linear increase in response beyond accelerations of 2.5 $m \cdot s^{-2}$ to about 4 $m \cdot s^{-2}$. This was determined by bench tests in the laboratory. It is not surprising, therefore, that laboratory treadmill studies have shown various brands of pedometers to be inaccurate for distances walked at slower speeds and in fast walking or running (Anderson, Masironi, Rutenfranz, & Seliger, 1978; Kemper & Verschuur, 1977; Saris & Binkhorst, 1977a; Washburn, Chin & Montoye, 1980). In field tests the same results occurred (Bassey et al., 1987; Saris & Binkhorst, 1977b; Washburn et al., 1980).

The same pedometer will give different results when worn by different individuals. Many people walk with greater impact on one foot, and their pedometer readings vary depending on which side the instrument is worn (Bassey et al., 1987; Gayle et al., 1977), although this has not been a universal finding (Verschuur & Kemper, 1980).

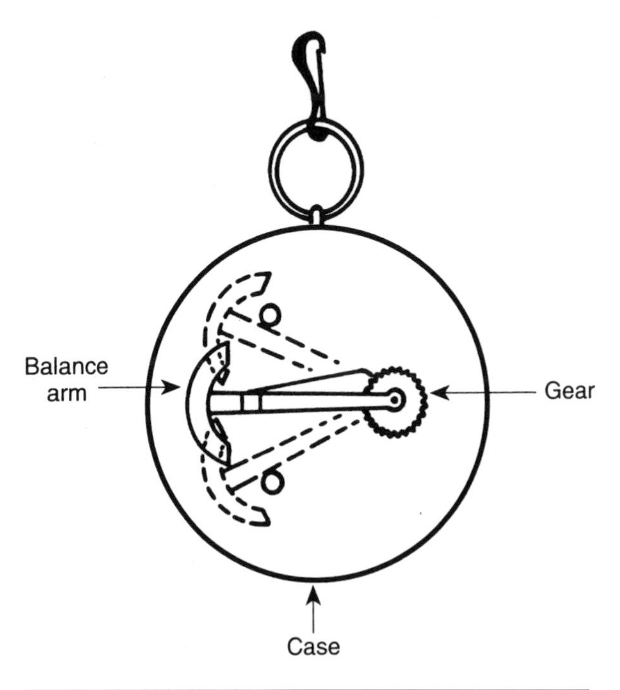

Figure 7.2 A diagram of the pedometer principle. From Kemper, H.C.G., & Verschuur, R. (1977). Validity and reliability of pedometers in habitual physical activity research. *European Journal of Applied Physiology*, **37**, 71-82. Reproduced with permission.

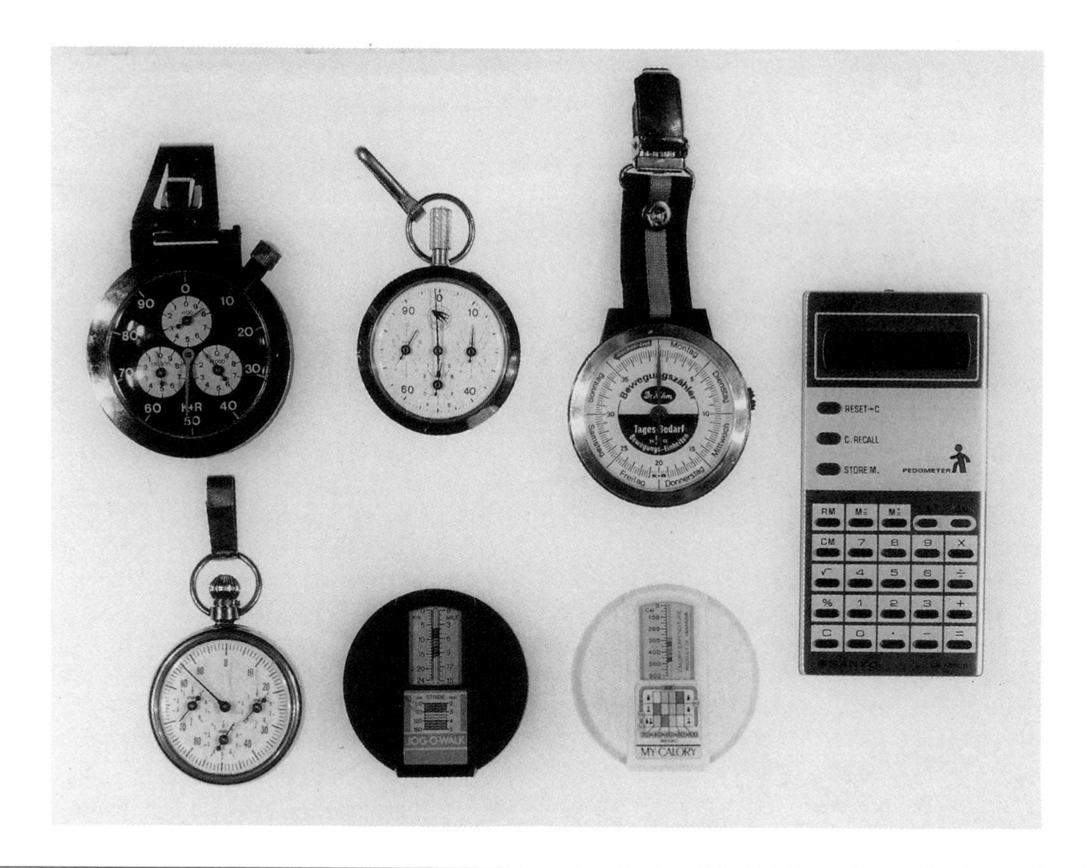

Figure 7.3 Various pedometers manufactured in Germany, Russia, and Japan.

Some brands of pedometers are more accurate than others. In controlled walking on a treadmill, the average error was ±5% for one brand, ±13% for another (Gayle et al., 1977). One problem with pedometers is that the tension of the springs varies, even in pedometers of the same brand. The tension should thus be adjusted and each instrument calibrated to give similar results (Bassey et al., 1987; Saris & Binkhorst, 1977a; Verschuur & Kemper, 1980). Verschuur and Kemper reduced the sensitivity of pedometer springs (Figure 7.4) so they would record only during running. This proved to be successful for their purpose, as is illustrated in Figure 7.5.

There have been additional attempts to validate the pedometer for estimating physical activity or energy expenditure; they are summarized in Table 7.1. Except for the study by Verschuur, Kemper, and Storm-van Essen (1987), these were field studies of total exercise or energy expenditure. When physical activity was measured by a pedometer, obese subjects were more active in one study (Chirico & Stunkard, 1960) but not in another (Wilkinson, Parkin, Pearlson, Strong, & Sykes, 1977). In walking or running, ankle-mounted pedometers give poorer results than waist-mounted pedometers (Saris & Binkhorst, 1977a; Washburn et al., 1980). However, in an activity like bicycle riding, an ankle-mounted pedometer would be more sensitive to leg movements.

In a laboratory bench test, the Yamasa pedometer was found to be reliable (reproducible; Bassey et al., 1987). However, test-retest correlation coefficients of various brands in treadmill walking at various speeds with the pedometer at the waist averaged from .49 to .70. In the same study, these coefficients averaged .60 to .62 for walking or running on the track (Washburn et al., 1980). In a measure of total activity, 30 employed men wore pedometers on their ankles and at the waist about 12 hr a day for a week; the day-to-day test reliability coefficients averaged .40 and .36, respectively (Gretebeck & Montoye, 1992). On weekdays only, these values were .53 and .48. With this rather poor reproducibility, it was estimated that an average of 5 or 6 days would be needed to reduce the error to about 5% (Gretebeck & Montoye, 1992). In this last study, weekday scores were greater than weekend scores, which is understandably opposite to the situation found in children (Verschuur et al., 1987). In any case, to estimate total activity, both weekdays and weekends must be sampled.

Irving and Patrick (1982) concluded, despite the imprecision of the pedometers, that the instruments measured gross differences in activity in five occupational groups. Others concluded that the pedometer was useful in reflecting the increased activity of a walking program in 108 factory workers (Bassey, Blecher, Fentem, &

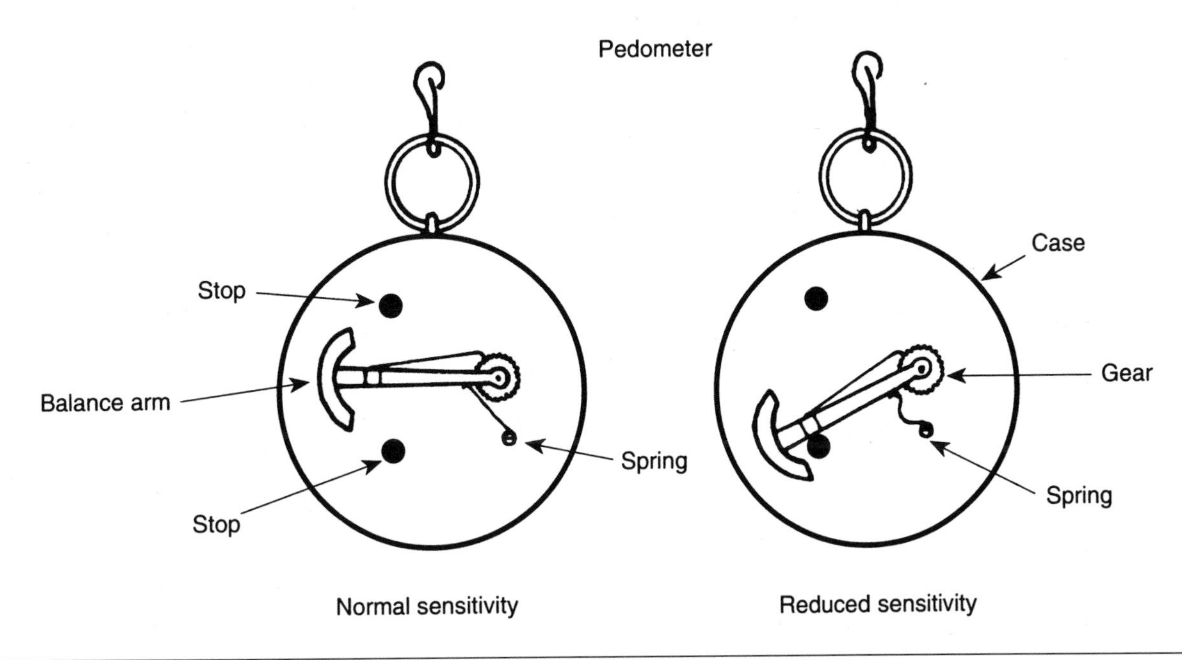

Figure 7.4 Settings for a pedometer to be used at normal and reduced sensitivity. From Verschuur, R., & Kemper, H.C.G. (1980). Adjustment of pedometers to make them more valid in assessing running. *International Journal of Sports Medicine*, **1**, 95-97. Reproduced with permission of the publisher, Georg Thieme Verlag, Stuttgart-New York.

Patrick, 1982). Kemper et al. (1976) used pedometers to estimate any change in physical activity as a result of adding two school physical education classes per week for 12- and 13-year-old boys. The study, carried out over an entire year, recorded activity both during school and during leisure time. As a result of adding the extra classes, the individual total weekly physical activity increased from 2% to 25%. But, as we shall see, other monitors give better results than the pedometer for estimating total activity in the field.

Other Step Counters

Step counters, which could also be labeled pedometers, have been developed that operate in a different way than those described in the previous section. To our knowledge, none of these is produced commercially. In 1967, Herron and Ramsden described an electronic device called a *telepedometer* that could be mounted in the heel of a shoe. The output indicating a step was taken could be transmitted up to 100 ft to a recorder. Running could be distinguished from walking.

Marsden and Montgomery (1972) developed a foot switch that fit into a shoe insert activated when about 75% of the person's weight was applied. Wires from the switch led to a counter weighing about 4 oz and measuring 1 by 1.6 by 2.5 in. that the subject carried in a pocket. A similar shoe insert was described by Barber, Evans, Fentem, & Wilson (1973), the output of which was recorded by a miniature tape recorder. The device can be set to record whether the subject is walking, lying, sitting, or standing. Members of this British group used this step counter in association with heart rate recording (Bassey et al., 1978; Bassey, Bryant, Fentem, MacDonald, & Patrick, 1980) and to validate pedometers (Irving & Patrick, 1982; Bassey et al., 1987).

More elaborate devices have been constructed for insertion into shoe heels in which force is recorded to monitor loads held, lifted, or carried. One of these used a strain gauge, mounted in a wooden shoe, that is recorded on a small Medilog tape recorder carried by the subject (Hagg, 1982). Anther such device uses an ultrathin capacitive force transducer inserted into a 2-mm-thick sole that can be worn in any type of shoe (Dion, Fouillot, & LeBlanc, 1982). The FM signal is transmitted to a receiver and Medilog tape recorder carried at the subject's waist. Similar devices have been used to record the forces on prostheses (Lovely, Berne, Solomonides, & Paul, 1982). Electronic step counters, like pedometers, are useful only for walking or running activities, but the force transducers in the shoe have wider application in reflecting the load carried or force exerted. However, there are disadvantages, too, including higher cost, the possible need for different inserts depending on shoe size or type, and the possibility of broken wires.

Motion Counters

Devices have been developed to count not only steps but also more general movements of the trunk or limbs.

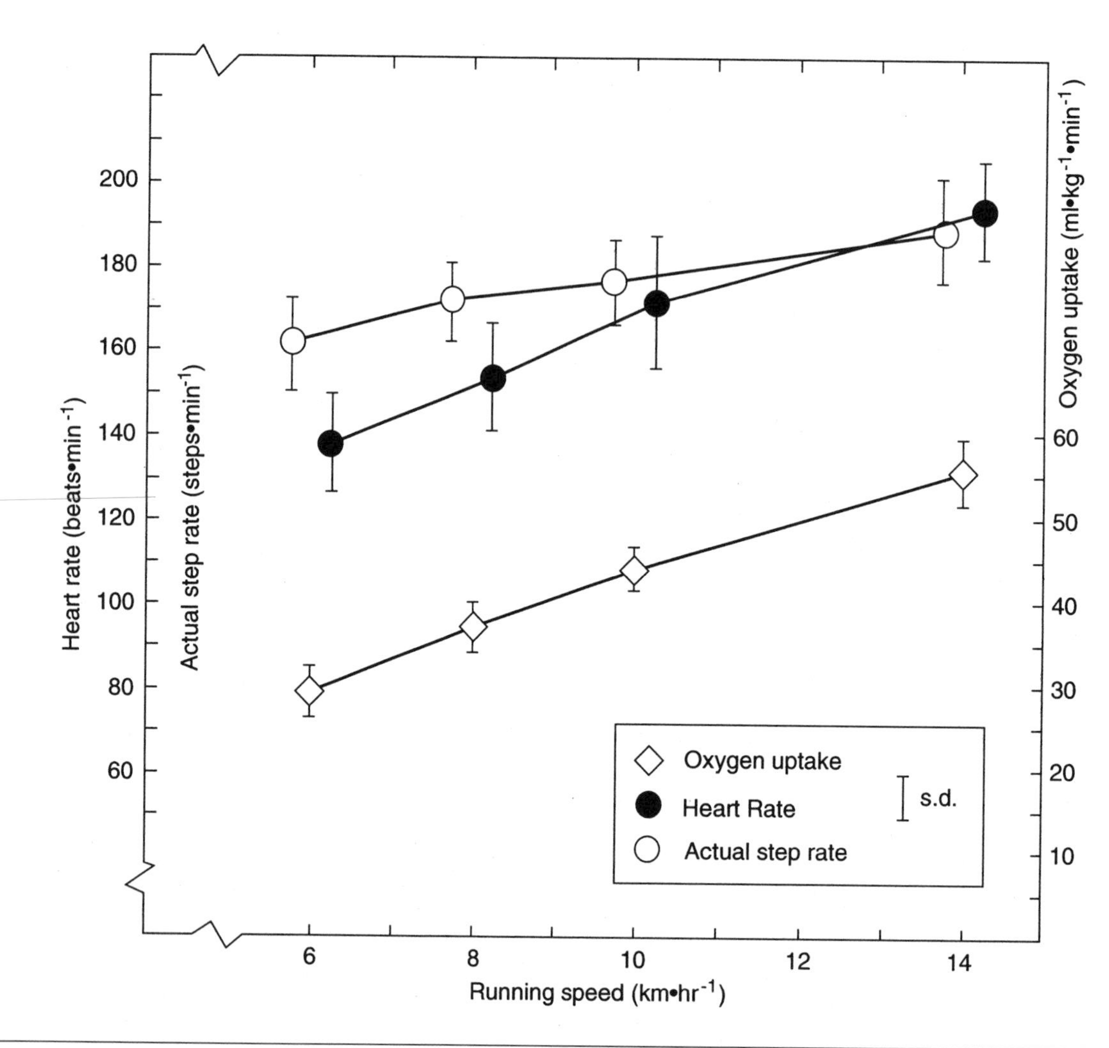

Figure 7.5 The relationship between running speed and O_2 and between heart rate and step rate after the sensitivity of the pedometer is reduced. From Verschuur, R., & Kemper, H.C.G. (1980). Adjustment of pedometers to make them more valid in assessing running. *International Journal of Sports Medicine*, **1**, 95-97. Reproduced with permission of the publisher, Georg Thieme Verlag, Stuttgart-New York.

For example, a sophisticated electronic counting device was developed by Colburn, Smith, Guarini, and Simmons (1976) at the National Institutes of Health. The transducer is a piezoelectric bilaminar bender element in a cantilever mounting. When the body or limb is accelerated sufficiently, a pulse is generated. These pulses are counted and stored during a given period (e.g., 10 min). The instrument can store 256 such period counts. Examples of data on several subjects are given, but no information is provided concerning the reliability or validity of the instrument.

Multiple Switches

In 1977, Schulman, Stevens, and Kupst described their *biomotometer,* which contained three mercury microswitches mounted at the waist in a plane parallel to the ground and at 120 degrees to each other. The unit measures 3-1/4 by 2-1/8 by 1-1/8 in. and weighs 3 oz. One problem would seem to be the possible double or triple counting from the various switches when activity is involved; but the authors report good reliability with four subjects (r = .84). The validity of the instrument was measured by a comparison with two actometers (to be described later in this chapter) in 20 children under various conditions. The correlation coefficients between the actometer and biomotometer readings were generally not very high. The average with the actometer mounted on the wrist was .62; mounted on the ankle, .60; and with the two actometers combined, .70.

A motor sensor has been devised (Taylor et al., 1980, 1982) that is attached to the lateral thigh by an elastic band or tape; it consists of six liquid mercury switches aligned on the faces of a cube. The sensor is connected to a microcomputer, the Vitalog MC-2 (Vitalog Corp., Redwood City, CA, USA), worn on the belt. The Vitalog

Table 7.1 Validation Studies for Estimating Physical Activity or Energy Expenditure With a Pedometer

Criterion	Population[a]	Correlation	Reference
Activity questionnaire	71 boys, ages 12-13 (2 days)	0.50*	Kemper & Verschuur, 1974
Energy expenditure estimated from individual calibrated heart rate	23 adults (1 day)	0.64*	Kashiwazaki, Inaoka, Suzuki, & Knodo, 1986
Endurance run	52 boys, mean age 12.5 yr	Not significant	Verschuur, Kemper, & Storm-Van Essen, 1987
W_{170}, bicycle ergometer/kg	(3 days, 2 seasons)	0.37*	
Endurance run	Same population, 4 yr later	Not significant	
W_{170}, bicycle ergometer/kg		Not significant	
Observation	11 children, ages 4.8 to 6.1 yr (5 days)	.93*	Saris & Binkhorst, 1977b
Actometer score[b] (sum of 2)		.95*	
Questionnaire/interview	102 boys, 131 girls	0.20	Kemper, 1992
Heart rate	(2 weekdays + 2 weekends, 2 seasons)	0.17	

[a]Number of days of pedometer recording shown in parentheses.
[b]Discussed in this chapter.
*Statistically significant, $p < .05$.

measures 4 by 8 by 12 cm and is programmed to store eight levels of physical activity representing a mean number of activations of the mercury switch over a predetermined period (e.g., 1 or 5 min). Berkowitz et al. (1985) used the device to study the relationship of physical activity and body fatness in 4- to 8-year-old children. Although they did not study the validity of the instrument, three monitors were very consistent in recording movement in more than 20 trials of a subject walking on a treadmill at 2 mph and 3% grade.

The developers of the Minilogger, a movement counter utilizing two mercury switches, claim it will record activity in any plane. The counter weighs about 100 g and is marketed by Minimibber Co., PO Box 3386, Sun River, OR, USA 97707. Its validity or reliability has not been studied. The unit is priced at $1,250, and the accompanying software is $250.

Large-Scale Integrated Motor Activity Monitor

One group of investigators developed telemetry equipment to record movements of psychiatric patients (Foster et al., 1972; Foster & Kupfer, 1973, 1975a, 1975b; Foster, McPartland, & Kupfer, 1977; Kupfer, Detre, Foster, Tucker, & DelGado, 1972; Kupfer & Foster, 1973; Kupfer, Weiss, Foster, Detre, DelGado, & McPartland, 1974; Weiss, Kupfer, Foster & DelGado, 1974a, 1974b). The transmitter, which resembled a wristwatch, led to a self-contained counter (McPartland, Foster, & Kupfer, 1976; McPartland, Kupfer, & Foster, 1976) and finally a simple device that used a mercury switch sensitive to a 3-degree tilt in a single axis (Foster, McPartland, & Kupfer, 1978a, 1978b; LaPorte et al., 1979). The unit, called a Large-Scale Integrated Motor Activity Monitor (LSI), is slightly larger than a wristwatch and may be worn on the trunk or limb. The present cost ($250) is prohibitive for some applications. The monitor is available commercially from GMM Electronics, Verona, PA, USA. A mechanical shaker was used in preliminary validation studies and the LSI counts were accurate. The counts also reflected the speed at which seven subjects walked on the treadmill and the distance two subjects walked on a track (Foster, McPartland, & Kupfer, 1978a). Since then, there have been a number of laboratory and field validation studies. Laboratory studies are summarized in Table 7.2 and field studies in Table 7.3.

The LSI 24-hr counts did not correlate with the coronary risk factor of high-density lipoprotein (HDL) cholesterol in 72 postmenopausal women (LaPorte et al., 1982). There also was almost no correlation between the LSI score and various HDL fractions in another sample of 246 postmenopausal women (Cauley et al., 1986) or in 35 postal carriers (Cook, Washburn, LaPorte, & Traven, 1985). However, the LSI results could differentiate between adult activity groups (Cauley et al., 1984) and between 10 active college physical education majors and 10 less active controls (LaPorte et al., 1979) and between three groups of monkeys classified as active, moderately active, or inactive (LaPorte et al., 1982).

The reproducibility of the LSI has been studied in several populations. Washburn and LaPorte (1988) had

Table 7.2 Validation of the Large-Scale Integrated Motor Activity Monitor: Laboratory or Controlled Studies

Criterion	Population	Location of monitor	Correlation or accuracy	Reference
$\dot{V}O_2$ daily activities	21 male adults	Wrist	$r = .60$; $SE_{est} = 7.9$ ml · kg^{-1} · min^{-1}	Montoye et al., 1983
		Trunk	$r = .36$; $SE_{est} = 9.2$ ml · kg^{-1} · min^{-1}	
$\dot{V}O_2$ cycling at various speeds	5 adults (20 measurements)	Ankle	$r = .46$; $SE_{est} = 5.9$ ml · kg^{-1} · min^{-1}	Hunter et al., 1989
Observation	14 children, ages 2-4	Trunk (90 min)	.90*	Klesges et al., 1984
Observation	25 male adults	Trunk (1 hr)	.65*	
Caltrac accelerometer[a]	25 female adults		.83*	Klesges, Klesges, Swenson, & Pheley, 1985
Observation	30 children, mean age 4 yr	Trunk (1 hr)	.40*	
Caltrac accelerometer[a]			.35*	
Walking speed	17 adults	Hip	.53*	Washburn & LaPorte, 1988
		Back	.38*	
Caltrac accelerometer[a]		Hip	.57*	
		Back	.38	

[a]Described in this chapter.
*Statistically significant, $p < .05$.

Table 7.3 Validation of the Large-Scale Integrated Motor Activity Monitor: Field Studies

Criterion	Population	Duration of monitor	Location of monitor	Correlation coefficient	Reference
Paffenbarger questionnaire	130 female adults on walking program	3 days	Trunk	.33* (.17*)	Cauley, LaPorte, Sandler, Schramm, & Kriska, 1987
Modified Paffenbarger questionnaire				.04 (.04)	
Baecke questionnaire	125 controls (coefficients in parentheses)				
Work index				−0.11 (.09)	
Leisure time index				.20* (.16*)	
Sport index				.17* (.07)	
Activity diary	10 college PE majors	2 days	Trunk	.69*	LaPorte et al., 1979
	10 college nonmajors		Ankle	.43	
Minnesota leisure time questionnaire			Trunk	.05	
			Ankle	.26	
Minnesota leisure time questionnaire	22 boys, ages 12-14 yr	2 days		.02	LaPorte et al., 1982
$\dot{V}O_2$max				−0.16	
Minnesota leisure time questionnaire	42 adult males	3 days		.11	
Paffenbarger questionnaire	72 adult females	3 days		.23*	
Caltrac accelerometer[a]	35 postal carriers	6 days	Trunk	.75*	Washburn et al., 1989
				.45*	
$\dot{V}O_2$max	18 children, age 6-8 yr			.59*	Fenster, Freedson, Washburn, & Ellison, 1991

[a]Described in this chapter.
*Statistically significant, $p < 0.05$.

17 adults walk twice at two speeds over a measured distance of 0.55 mile. Test-retest correlation coefficients were low—.22 to .47. In the study of four subjects performing a number of different daily activities, test-retest coefficients were slightly better—$r = .74$ for wrist-mounted and .63 for waist-mounted (Montoye et al., 1983). In the study of 33 postal carriers (Washburn, Cook, & LaPorte, 1989), Day 1 versus Day 2 produced a correlation of .46, and 3 months later the test-retest correlation was .56. The original investigation of the reliability of the LSI (Foster et al, 1978a) resulted in higher test-retest coefficients (.76 to .99); seven subjects were retested in walking and running at various speeds in a controlled laboratory setting and with the LSI worn on the ankle, wrist, or trunk.

The LSI has at least two advantages over pedometers. First, the standardization of instruments should be better because mechanical springs are not used. Second, it may be useful in a greater number of activities because it operates on tilt rather than impact. However, some of the limitations of pedometers also apply to the LSI. Specific days sampled may not reflect year-round activities. The LSI counts movement as does the pedometer and hence does not reflect forces and energy expenditure well. Also, the LSI does not hold up well under hard use, and is difficult to repair.

Portable Accelerometers

When a person moves, the limbs and body are accelerated, theoretically in proportion to the muscular forces responsible for the accelerations and thus to energy expenditure. Accelerations have been measured in the laboratory with high-speed photography, force plates, and subject-mounted accelerometers. For the most part, these instruments have no direct application to the measurement of habitual physical activity in the field. Nevertheless, we present a brief review because the instrumentation is important and because the studies illustrate the theoretical limitations of the approach (Cavagna & Margaria, 1966; Cavagna, Saibene, & Margaria, 1964; Williams, 1985).

Theoretical Considerations

Portable (i.e., subject-mounted) single plane accelerometers have been used to study limb movements in the laboratory with the signal telemetered (Dewhurst, 1977) or recorded directly (Josenhans, 1967). Force plate measurements are moderately correlated with energy expenditure in particular movements (Brouha, 1960; Brouha & Smith, 1958; Montoye, Servais, & Webster, 1986). There is also good agreement between measurements from a subject-mounted accelerometer and high-speed photographic records when a subject is walking (Cavagna, Saibene, & Margaria, 1961, 1963).

Cotes and Meade (1960) measured the vertical oscillations of the trunk that occur in walking by means of a wire attached to a firmly fitting waist belt and connected to a mechanical integrator. They found that energy expenditure was fairly closely related to vertical lift work (vertical lift per step × step frequency × body weight). Reswick, Perry, and Antonelli (1979) used photographs and a head-mounted accelerometer to show that during natural walking, lift per step (and hence lift power) correlated well with the integral of absolute vertical acceleration versus time. They also found that for walking subjects, the integral of absolute vertical acceleration versus time could predict $\dot{V}O_2$.

Bhattacharya, McCutcheon, Shvartz, and Greenleaf (1980) taped vertically mounted linear accelerometers on the ankle, back, and forehead. The signals were telemetered to a recorder. Subjects walked or ran on a treadmill at four speeds and jumped on a trampoline to four different heights while analog acceleration tracings, heart rate, and $\dot{V}O_2$ were recorded. Peak accelerations and $\dot{V}O_2$ were normalized for body weight. Although the authors acknowledged that normalized peak accelerations represent a major component of external work, the accelerations were greater for trampoline jumping than walking or running at the same $\dot{V}O_2$. In each activity, peak accelerations and $\dot{V}O_2$ increased as the intensity increased.

Although the instruments in these laboratory investigations are not suitable for epidemiologic studies or patient monitoring, the results reinforce the notion that measuring acceleration of body mass might be useful in estimating energy expenditure. Portable accelerometers have also been used to study gait or movements associated with gait. Examples of this kind of investigation are studies by Saunders, Inman, and Eberhart (1953) on pathological gait and by Morris (1973) measuring acceleration and deceleration of the limbs.

It is clear from these investigations that even everyday activities such as walking, bending, and climbing stairs are complex movements. All of the energy expended is not reflected in acceleration or deceleration of the body mass. In walking, for example, some counterbalancing movements of the limbs that require energy may not contribute to acceleration or deceleration of the body. Rotational acceleration of the limbs in walking or running continuously alters the center of gravity of the total body. Furthermore, with increasing running speed, stride length is increased while step frequency remains relatively constant (Saris & Binkhorst, 1977a).

These changes in movement pattern will have a different effect on accelerometer readings compared to the linear increase in energy expenditure. Isometric muscle contraction, as in fixing parts of the skeleton, utilizes energy without being reflected in movement. Eccentric contraction of the thigh muscles in bending the legs, which requires energy expenditure, more than likely reduces the recorded deceleration of the body. Carrying a backpack or other weights adds to the energy cost without increasing the acceleration of body mass (Taylor, Heglund, McMahon, & Looney, 1980). In activities like bicycling and rowing, an accelerometer attached to the waist will record little movement even though energy expenditure may be high.

On the other hand, in classifying individuals on the basis of habitual physical activity (or energy expenditure during exercise), we are dealing with a wide range, from people who are extremely sedentary to those whose lives encompass considerable occupational or leisure activity, or both. In patient populations, one might be concerned with an intervention program that adds considerable physical activity to an otherwise sedentary lifestyle. Thus, perhaps in applications such as these, finer biomechanical analyses and problems assume less importance in the development of movement assessment instrumentation. In fact, in some movements, such as horizontal walking, total energy expenditure is reflected in the number of repetitive movements (steps), and rate of energy expenditure is roughly proportional to the number of such cyclic movements per unit of time (Montoye et al., 1983). A simple movement counter may thus estimate energy expenditure as well as does a more elaborate device. However, in more complicated movements, such as playing tennis or basketball, gardening, walking up and down stairs, cycling, and doing housework, a more complicated instrument may be more useful. This has led to the development of portable accelerometers for use in the field.

The Actometer

Schulman and Reisman (1959), interested in hyperactivity in children, used a modified self-winding wristwatch (an Omega calendar watch with a number of the internal parts removed) to record acceleration and deceleration. Later studies by others indicated that less expensive watches of the calendar type worked as well (Groenewegen, 1979). Although data on the reliability and validity of the wristwatch device worn in the field have received limited attention, Schulman and Reisman have tested both on a mechanical apparatus and state that they are accurate. Their device, called an actometer, is designed to measure acceleration and deceleration in one plane.

Reproducibility (Reliability). Bench tests in the laboratory have revealed that when the same actometer is tested repeatedly, results are reproducible. However, there are large differences between actometers (Avons, Garthwaite, Davies, Murgatroyd, & James 1987; Eaton & Keats, 1982; Johnson, 1971; Saris & Binkhorst, 1977a; Sweetman, Edwards, & Anderson, 1978). The same problem is encountered as with pedometers—the tension on the spring varies. Most investigators have developed a correction factor for each actometer in an effort to make the results comparable.

There have also been attempts to measure the reliability (reproducibility) of the actometer when worn by a subject. Any inconstancy in the score under these circumstances reflects not only the unreliability of the instrument but the inconstancy of the activity itself. These data indicate how many repetitions are needed for a stable estimate of physical activity. When one actometer was worn on the wrist and one on the ankle for 1-1/2 to 2 hr and the two values were summed, the test-retest coefficient with 11 children aged 5 and 6 was .52 (Loo & Wenar, 1971). Maccoby, Dowley, Hagen, and Dagerman (1965) also found a low reliability of the instruments in classroom sessions.

If several recordings are averaged, the reproducibility of the average is much better. For example, Bell, Weller, and Waldrop (1971) averaged recordings of children's actometer scores on 3 days and correlated the mean with the mean of 3 other days; the correlation was .68. Similarly, Halverson and Waldrop (1973) did an analysis with 58 preschool children. The two means, each based on 3 days of observation, were correlated at .81 for boys and .92 for girls. With 27 children, mean age about 4 years, who wore a wrist actometer during free play, a low reliability was reported for 1-day recordings of only .33, whereas this value rose to .90 with 14 days of recording (Eaton, 1983). Much the same observations were made by Halverson and Post-Gorden (1984). With 40 young children during indoor free play, a 1-day reproducibility coefficient was only .17 but rose to .67 for 10 days and .88 for 20 days. These authors concluded that at least six or seven observations are necessary to achieve reasonable reliability.

A similar instrument manufactured by Timex produced a test-retest coefficient of .80 when 5 days' average was correlated with a 5-day average the following week. Thirty-three children, aged 6 to 16, served as subjects (Massey, Lieberman, & Batarseh, 1971). Actometer scores (average of three observations) of 129 three-year-old children showed some stability in that these means were correlated with an average of four measurements a year later, producing coefficients of .44 for boys and .43 for girls (Buss, Block, & Block, 1980). Actometers were worn on the wrist. Fourteen adult patients wore two actometers on the same limb for one day in a study by Morrell and Keefe (1988). The correlation coefficient comparing the output of the two actometers was .997. However, when the patients wore one actometer and walked assigned distances on 3 days, the actometer scores were poorly correlated between days.

Validity. Laboratory studies included a report by Saris and Binkhorst (1977a) in which the scores of an actometer mounted on the ankle correlated well with energy cost of walking and running at various speeds. The correlation was better than with waist pedometer scores. In another laboratory study (Avons et al., 1987) 12 males, mean age 23, spent 17-1/2 hr in a metabolic chamber, during which time they performed various standardized activities and wore one actometer on the wrist, one on the ankle, and one at the waist. Correlation coefficients reflecting the association of actometer scores and oxygen uptake (energy expenditure) were .75 (wrist), .89 (waist) and .96 (ankle), all of which are statistically significant. Finally, $\dot{V}O_2max$ was correlated with 24-hr actometer scores for 8- and 13-year-old children (Sunnegårdh & Bratteby, 1987). The coefficients ranged from .41 to .48 (all statistically significant) for boys and girls aged 13 and boys aged 8. The coefficient for the 8-year-old girls was not significant. Field studies are summarized in Table 7.4.

The Caltrac

Dawson (1959) some years ago used a phonograph cartridge as a transducer to measure movement. A small, electronic single-plane accelerometer, which might be useful in population studies, has been tested in one of our laboratories. The first prototype of this instrument also used a phonograph cartridge as the accelerometer transducer (Wong, Webster, Montoye, & Washburn, 1981), and the output was read on a Curtis meter. The instrument was tested by having 21 subjects walk and run on a motor-driven treadmill at various speeds and grades and perform bench stepping, knee bending, and floor touching at various speeds while wearing the accelerometer at the waist and an LSI mercury switch at the waist and on the wrist. $\dot{V}O_2$ was measured simultaneously. When the results of the various activities were pooled, the accelerometer estimated $\dot{V}O_2$ better than the mercury switches (Montoye et al., 1983). The standard error of estimate for the accelerometer was 6.6 ml $O_2 \cdot kg^{-1} \cdot min^{-1}$; for the mercury switches it was 7.9 and 9.2 when mounted on the wrist and waist, respectively.

In a second prototype, the phonograph cartridge was replaced by a piezoelectric bender element, a digital output replaced the Curtis meter, power consumption and size were reduced, and a low-battery indicator and reset capabilities were added (Servais, Webster, & Montoye, 1984). The output of the accelerometer worn at the waist was almost identical to the vertical component of a force platform while subjects did bench stepping and knee bends (Servais, Webster, & Montoye; Figure 7.6). However, when the subject repeatedly bent over to touch the floor, agreement was poor, probably because the portable accelerometer no longer is maintained in a vertical position on such movements. This finding points to an advantage of a triaxial accelerometer. This second prototype was also tested in bicycling at various speeds with the device worn on the thigh or anklê. Again, $\dot{V}O_2$ was measured and was estimated quite well by the accelerometer (Hunter et al., 1989). The standard error of estimate was 3.3 ml $O_2 \cdot kg^{-1} \cdot min^{-1}$ ($r = .85$). The LSI resulted in a larger standard error of estimate (5.9) and lower correlation (.46).

At this point, the portable accelerometer was modified again and produced commercially, sold under the name Caltrac. It is designed to clip firmly to a belt at the waist. It is important that the accelerometer fit tightly to the body if accelerations and deceleration of body mass are to be accurately estimated (Servais et al., 1984). The commercial version provides for the entry of age, weight, height, and gender so that the readout is in kilocalories. After these data are entered for a subject, resting metabolic rate is recorded if the Caltrac is not moved. If movement of the subject occurs, its metabolic cost is added to the resting rate. The Caltrac can also be used as a movement counter by programming the following data into the unit: weight, 85 lb; height, 8 in.; age, 99 years; gender, male. The output is then in counts rather than kilocalories. The transducer, a piezoelectric bender element, is made of two layers of piezoceramic material with a brass center layer. When the body accelerates, the transducer, which is mounted in a cantilever beam position, bends, and a charge is produced proportional to the force exerted. The area under this acceleration-deceleration wave form is integrated and summed.

Reproducibility. The reproducibility of the instruments can be assessed by repeating the measurements in a bench test or other carefully controlled activity with humans. Another approach is to mount two units on a subject and compare their scores. Studies in which subjects wore two Caltracs are summarized in Table 7.5.

Test-retest comparisons under controlled conditions are shown in Table 7.6. Reproducibility of the Caltrac has also been studied under field conditions (see Table 7.7). In the study by Gretebeck and Montoye (1992), if an estimate for all 7 days of the week is wanted, 3 days for kilocalories and 6 for METs recording, including at least one weekend day, are needed. More days are needed if the score is expressed in METs because the influence of body weight is essentially eliminated, which is not true when the score in in kilocalories.

Validity. A number of Caltrac validation studies besides the initial evaluation have been conducted in the

Table 7.4 Actometer Scores Compared to Other Methods of Assessing Physical Activity: Field Studies

Other method	Population	Location of actometer	Correlation	Reference
Teacher questionnaire	11 children, ages 4-6 yr	Ankle and wrist (summed)	Differentiated most and least active	Saris & Binkhorst, 1977b
Observation		Ankle	$r = .97$*	
		Wrist	$r = .71$*	
		Sum	$r = .97$*	
Diary and regression equations	6 male adults	Leg, waist, wrist	Poor	Avons et al., 1987
Teacher observation	129 children, ages 3 and 4 yr	Wrist	$r = .53$* to .61* (average of 3 or 4 days)	Buss et al., 1980
Teacher observation	58 preschoolers	—	$r = .82$* (average of 6 days)	Halverson & Waldrop, 1973
2 teachers' observations	140 children, ages 2 1/2 to 3 1/2 yr	Back	$r = .17$* (one day) $r = .46$* (6-day average) $r = .67$* (10-day average) $r = .88$* (20-day average)	Halverson & Post-Gorden, 1984
Teacher observation	16 children, mean age 4 yr	Wrist and ankle (summed)	$r = .33$ and .34	Eaton & Keats, 1982
Staff activity ratings	27 children, mean age about 4 yrs	Wrist	$r = .69$* (average of 14 days)	Eaton, 1983

*Statistically significant, $p < .05$.

laboratory or under controlled conditions (Montoye, et al., 1983). These are summarized in Table 7.8. Field studies correlating Caltrac results with other methods of assessing physical activity are shown in Tables 7.9 and 7.10.

Because the Caltrac does not contain springs in which tension can vary as do pedometers and the actometer, one would expect the interinstrument variation to be minimal; this is supported by the data in Table 7.5. Under controlled conditions (Table 7.6), the test-retest correlation, even in children, is good. However, in field conditions it is poorer, which is not simply a reflection of errors in the instrument but more likely of a change in the physical activity of the subjects. Under controlled conditions, energy expenditure is estimated well for horizontal walking, running, and cycling but poorly for grade running or for activities with a large nonvertical component (Table 7.8).

It is clear that for certain activities, such as cycling, the position of the Caltrac is critical for estimating energy expenditure. The poorer correlation under field conditions is likely due as much or more to error in the criterion measures. Also, in some studies only 1-day or 1-hr recordings of the Caltrac were used. Then, too, many of the questionnaires estimated year-round activity, whereas the Caltrac recordings were for a limited period. Clearly, in almost all situations, the Caltrac gives more valid results and the reproducibility is better than with the LSI. The quality control of the Caltrac has left something to be desired; about 5% to 10% of them malfunctioned. However, the unit has been redesigned to improve quality control; the new version is marketed by Muscle Dynamics Fitness Network, 20100 Hamilton Ave., Torrence, CA, USA 90502. A picture of the newer version of the single-plane Caltrac is shown in Figure 7.7.

The basic principles of the original Caltrac remain in the modified version. However, in addition to being more dependable, the new instrument is powered by two AAA batteries instead of the lithium batteries, and two additional modes have been added, one for cycling and one for weightlifting. It is still possible to obtain a readout in kilocalories if gender, weight, height, and age are programmed in either English or metric units. Activity-only kilocalories can be obtained if desired, instead of total kilocalories (resting kilocalories are automatically subtracted from total kilocalories). At this writing, the unit sells for $89 U.S. Included with the Caltrac are an instruction booklet, a demonstration video-tape, a booklet containing calorie contents of various foods, and a booklet on health and exercise. There is

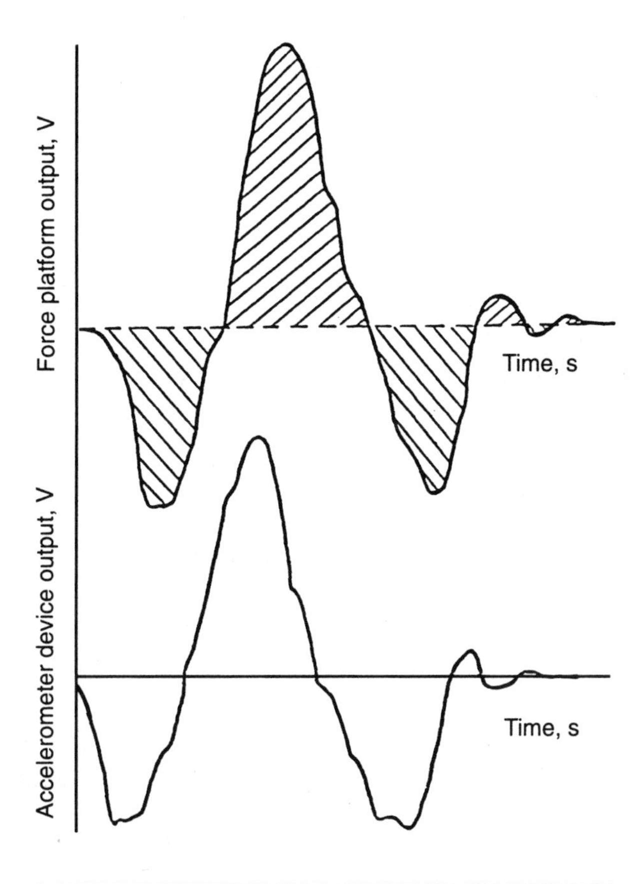

Figure 7.6 The output of the vertical component of a force platform (top) and of a portable vertical accelerometer (bottom). From Servais, S.B., Webster, J.G., & Montoye, H.J. (1984). Estimating human energy expenditure using an accelerometer device. *Journal of Clinical Engineering*, **9**, 159-171. Reproduced with permission from: Journal of Clinical Engineering copyright © 1984, Quest Publishing Company, 1351 Titan Way, Brea, CA 92621, (714) 738-6400.

not yet any published research on the new Caltrac, particularly regarding the cycling and weightlifting modes. The instrument would appear to be very useful in estimating increases in physical activity when a walking or jogging program is instituted.

Other Single-Plane Accelerometers

As far as we know, none of the accelerometers discussed in this section is available commercially. We describe them briefly for the reader's information and, in some instances, to provide data on the validity of accelerometers for estimating energy expenditure or physical activity. Colburn, Smith, Guarini, and Simmons in 1976 used a piezoelectric transducer to measure acceleration. Continuous acceleration-deceleration is not recorded because a pulse is stored only when sufficient acceleration is experienced. Kripke, Mullaney, Wyborne, and Messin (1978) reported the development of a portable accelerometer worn on the wrist to measure activity in psychiatric patients. The transducer was a piezoelectric element, with output recorded on a Medilog recorder that the subject wore on a belt.

Dewhurst (1977) designed a single-plane accelerometer that could be worn on the wrist or ankle. The output is transmitted to a receiver. Circuit diagrams are provided for this telemetry system. In 1980, Lipsey received U.S. Patent No. 4,192,000 for an accelerometer called an *electronic calorie counter*. Like the device developed by Colburn et al. (1976), continuous acceleration is not recorded. An electronic gate records impulses when the acceleration exceeds a certain minimum; the impulses are accumulated and used to estimate caloric expenditure. As far as we know, this device also has not been put into production and no validity data are available.

Table 7.5 Interinstrument Comparison of the Caltrac Accelerometer Under Controlled Laboratory Conditions: Two Caltracs Worn Simultaneously

Population and conditions	Location on body	Reliability	Reference
4 male adults, 14 activities (N = 56 points)	Waist	r = .94*	Montoye et al., 1983
20 adults walked on treadmill at 3.2-6.4 km · hr^{-1}	Waist	r = .94*, difference in means 9 kcal/15 min	Pambianco, Wing, & Robertson, 1990
15 children, mean age 10.8 yr (40 data points)	Waist	r = .89*	Sallis et al., 1990; Sallis, Buono, Roby, Carlson, & Nelson, 1990

*Statistically significant, $p < .05$.

Table 7.6 Test–Retest Comparison of the Caltrac Under Controlled Conditions

Population and conditions	Location on body	Reliability	Reference
6 adults walking on treadmill 3 times over 2 weeks	Waist	No significant difference in means	Pambianco et al., 1990
18 children, mean age 10.4 yr 2 ten-min walk/run, treadmill	Unknown	$r = .79$*	Kastango & Freedson, 1991
17 adults, 2 trials	Waist, normal	$r = .70$*	Washburn & LaPorte, 1988
Normal and fast walks, 0.55 miles	Waist, fast	$r = .87$*	
	Back, normal	$r = .73$*	
	Back, fast	$r = .83$*	
17 children, ages 9-12 yr (2 days)	Waist and back	$r = .93$*	Broskoski, Pivarnik, & Morrow, 1991

*Statistically significant, $p < .05$.

Table 7.7 Reliability of the Waist-Mounted Caltrac in Field Studies

Population and conditions	Reliability	Reference
35 children, ages 8-13 yr	No significant difference in means, $r = .30$*	Sallis et al., 1990; Sallis, Buono, Roby, Carlson, & Nelson, 1990
23 third graders, 2 three-day periods	2 days explained 65.9% of variance	Simons-Morton & Huang, 1991
24 fifth graders, 3 days	4 days explained 92.0% of variance	
19 adult physical therapists, 2 work days	$r = .91$*	Balogun, Fanna, Fay, Rossman, & Pozyc, 1986
28 female adults, 4 days	No significant difference in means $r = .61$* to .85*	Mahoney & Freedson, 1990
31 college students, 3 days	$r = .35$* to .54* (mean .42)	Williams, Klesges, Hanson, & Eck, 1989
30 children, ages 5-9 yr, 3 days	r .38* to .79*	Freedson, 1991
35 male postal carriers, 2 days in same week	$r = .47$* and .62*	Washburn et al., 1989
2 days vs. 2 days, 3 months intervening	$r = .44$*	
72 adults, ages 20-59 yr, 2 days, 6-month interval	kcal, $r = .84$* METs, $r = .49$*	Jacobs, Ainsworth, Hartman, & Leon, 1993
44 young adults, estimated r for 6-day mean	Caltrac counts, $r = .86$*	Dishman, Darracott, & Lambert 1992
30 employed males		Gretebeck & Montoye, 1992
1 day vs. 1 day	kcal, $r = .72$* METs, $r = .34$*	
2 day vs. 2 day	kcal, $r = .84$* METs, $r = .75$* No significant difference in means	

*Statistically significant, $p < .05$.

Bhattacharya et al. (1980) reported on the use of portable accelerometers attached to the head, back, and ankle. In a laboratory study with eight young males, the peak acceleration was shown to be closely correlated with $\dot{V}O_2$ and heart rate during treadmill walking and running and trampoline jumping. The accelerometer signals were telemetered. Finally, Tamura, Nakajima, Togawa, Wakabayashi, and Osana (1985) developed a strain gauge accelerometer and provided a schematic for the instrument. They used the accelerometer with university

Table 7.8 Validation Studies of the Caltrac Under Laboratory or Controlled Conditions

Criterion	Population	Location on body	Correlation	Reference
Metabolic chamber, kcal/day	29 female adults	Waist	kcal, r = .92* Underestimated by 14%	Schutz, Froideraux, & Jequier, 1988
$\dot{V}O_2$ treadmill walk, 4 speeds	25 adults	Waist	Overestimated by 13-53% $\dot{V}O_2 \cdot kg^{-1} \cdot min^{-1}$, r = .76* $kcal \cdot kg^{-1} \cdot min^{-1}$, r = .91*	Balogun, Martin, & Clendenin, 1989
$\dot{V}O_2$ treadmill walk/run, 3 speeds	15 children, mean age 11 yr	Waist	Counts, r = .82*	Sallis et al., 1990; Sallis, Buono, Roby, Carlson, & Nelson, 1990
$\dot{V}O_2$ treadmill walk, 3 speeds	20 adults	Waist	kcal, r = .68* to .79* for various speeds Overestimated by 9-13%	Pambianco et al, 1990
$\dot{V}O_2$ treadmill walk/run	18 children, mean age 10 yr	—	r = .70*	Kastango & Freedson, 1991
$\dot{V}O_2$ treadmill walk/run, 7 speeds	11 subjects	Waist	Walking: overestimated kcal, r = .94* Counts, r = .89* Running kcal, r = .70* Counts, r = .24	Haymes & Byrnes, 1991
$\dot{V}O_2$ treadmill walk, 2 speeds	17 children, ages 9-12 yr	Waist Back	$kcal \cdot kg^{-1}$, r = .89 Overestimated by 5-46%	Broskoski et al., 1991
$\dot{V}O_2$ treadmill walk/run, 2 speeds, 2 grades	20 adults	Waist Back	Overestimated horizontal walk, Underestimated grade walk	Losen, Ryan, & Doigener, 1991
Walk, 0.55 miles, 2 speeds	17 adults	Waist Back	r = .81* with walk r = .90* speed	Washburn & LaPorte, 1988
$\dot{V}O_2$ bike riding, 4 speeds	20 adults	Ankle	$kcal \cdot kg^{-1}$, r = .85* Standard error of estimate, 3.3 $ml \cdot kg^{-1} \cdot min^{-1}$	Hunter et al., 1989
Basketball game, video observation, heart rate (calibrated)	20 high school students	Waist	kcal, r = .95* Heart rate kcal, r = .92*	Ballor, Burke, Knudson, Olson, & Montoye, 1989

*Statistically significant, $p < .05$.

students, schoolchildren, and preschool children. The output of the accelerometer was highly correlated with $\dot{V}O_2$ and heart rate during laboratory experiments. The exercises involved jumping at various frequencies and bench stepping.

Recently Janz (1994) described a single-plane accelerometer based on the same principle as the Caltrac. With 31 children as subjects, the day-to-day correlation of the accelerometer counts produced correlations from .32 to .53. The average correlation coefficient between the activity counts and daily net heart rate (heart rate above resting heart rate) was .57.

Triaxial Accelerometers

From the earlier discussion in this chapter it is clear that a single-plane accelerometer mounted at the waist

Table 7.9 Relationship of Caltrac Recordings With Other Methods of Assessing Physical Activity

Criterion	Population	Length of Caltrac recording	Correlation	Reference
$\dot{V}O_2$max	78 adults, ages 20-59 yr	14 two-days administrations	kcal, $r = -.20$* METs, $r = .20$*	Jacobs, Ainsworth, Hartman, & Leon, 1993
LSI[a]	35 postal carriers	6 days	kcal, $r = .75$*	Washburn et al., 1989
Observation	28 children, ages 2-4 yr	1 day (counts)	All day, $r = .54$* Hourly, $r = .54$* to .95*	Klesges & Klesges, 1987
Observation	50 adults, ages 18-41 yr	1 hr recreation	$r = .69$*	Klesges et al., 1985
LSI[a]			$r = .83$*	
Observation	30 children, ages 2-6 yr		$r = .40$*	
LSI[a]			$r = .42$*	
Video observation	48 preschool children	1 hr	$r = .86$*	Nolan, Danner, Dewalt, McFadden, & Kotchen, 1990
Doubly labeled water	8 female adults, mean age 65 yr	7 days	kcal, underestimated, 3.8% METs, underestimated, 3.8%	Gretebeck, Weisel, & Boileau, 1992
Doubly labeled water	28 employed males	7 days	kcal, underestimated, 4.1% METs, underestimated, 4.3% kcal, $r = .55$* METs, $r = .34$*	Gretebeck, Montoye, & Porter, 1991, 1993
Activity heart rate minus resting heart rate	33 children, ages 8-13 yr	1 day	Day 1, counts, $r = .54$* Day 2, counts, $r = .42$*	Sallis et al., 1990; Sallis, Buono, Roby, Carlson, & Nelson, 1990

[a]Large-Scale Integrated Motor Activity Monitor.
*Statistically significant, $p < .05$.

reflects vertical movement well, but is not very effective when the motion is in other than the vertical plane. This suggests that a triaxial (three-dimensional) accelerometer would estimate energy expenditure better when various kinds of exercise are involved.

Three-dimensional accelerometers have been developed for measuring tremor and ataxia in neurological patients (Frost, 1978; Riker, 1980), but these have not been validated for the measurement of total energy expenditure as a result of doing various kinds of activities. Also, even though the recorder and ancillary equipment may be carried in an attaché case, the system is not sufficiently portable for epidemiologic studies and is not available commercially.

Redmond and Hegge (1985) at Walter Reed Army Institute of Research developed a triaxial accelerometer that can be worn on the wrist. Commercially available miniature accelerometers were mounted in a wristwatch case. The output was transmitted via transducer leads to a four-channel miniature tape recorder (Oxford Medilog Model 4-24) worn on the subject's belt. The fourth channel is occupied by a time signal. When the tape recorder is removed from the subject, the analog activity signals are converted to digital form, which can be analyzed in a number of ways. This instrument is available commercially under the name Actigraph from Ambulatory Monitoring, 731 Saw Mill River Rd., Ardsley, NY, USA 10502. It weighs about 85 g and measures about 6 by 9 by 2 cm. The cost at present is prohibitive for many studies—about $1,750 for the recording unit worn by the subject and about $2,500 for the data recovery interface unit.

Before portable triaxial accelerometers were available commercially, several investigators studied the reliability and validity of simulated triaxial accelerometer devices. Halverson and Waldrop (1973) mounted three actometers (Bell, 1968) at right angles on 58 children (33 males, 25 females) for six 15-min periods throughout a 5-week session. Reproducibility was good when the accelerometer score (the sum of the three watches) averaged over Days 1, 3, and 5 was correlated with the average of Days 2, 4, and 6 ($r = .81$ for males, .92 for

Table 7.10 Correlation of Caltrac Results With Questionnaire or Diary Assessment of Physical Activity

Questionnaire or diary	Population	Length of Caltrac recording	Correlation	Reference
Questionnaire for the day	35 children, ages 8-13 yr	1 day	Day 1, r = .49* Day 2, r = .39*	Sallis et al., 1990; Sallis, Buono, Roby, Carlsson, & Nelson 1990
Paffenbarger questionnaire	35 postal carriers	6 days	kcal · kg^{-1}, r = .36*	Washburn et al., 1989
Baecke questionnaire	28 female adults	4 days	r = .53*	Mahoney & Freedson, 1990
Baecke questionnaire	30 male adults	7 days	kcal, r = .53* METs, r = .40*	Gretebeck & Montoye, 1990
Paffenbarger questionnaire			kcal, r = .87* METs, r = .69*	
Health Insurance Plan of New York questionnaire			kcal, r = .28 METs, r = .51*	
7-Day/Stanford questionnaire			kcal, r = .72* METs, r = .69*	
Minnesota leisure time and Tecumseh questionnaires			kcal, r = .86* METs, r = .63*	
4 checklists and questionnaires	69 children, 4th grade	3 days	METs, r = .11 to .36*	Roby, Sallis, Kolody, Condon, & Goggin, 1992
Diary	45 adults, ages 65-85 yr	3 days (counts)	% time standing, r = .28* % time sitting, lying, r = –.20 % time action, r = .28*	Washburn, Janney, & Fenster, 1990
7-Day/Standford questionnaire	7 males, 26 females	7 days	kcal, r = 79* (.79*)[a]	Miller, Freedson, & Kline, 1994
Baecke questionnaire			kcal, r = .32 (.40)[a]	
Godin-Shephard questionnaire			kcal, r = .37 (.45)[a]	
4 weeks of leisure time activity, 14 administrations	73 adults, ages 20-59 yrs	14 two-days administrations	r kcal/r METs	Jacobs et al., 1993
Lipid Research Clinics			–.01/.31*	
Godin leisure			–.20*/.21*	
Health Insurance Plan of New York			–.02/.32*	
Cardia (heavy activity)			.07/.14	
Baecke questionnaire			.06/.31*	
7-Day Recall (METs)			–.08/.19*	
Paffenbarger questionnaire			.12/.33*	
Minnesota heart leisure index			–.03/.30*	

(continued)

Table 7.10 *(continued)*

Questionnaire or diary	Population	Length of Caltrac recording	Correlation	Reference
Stanford usual activity			−.16/.28*	
Vigorous index			−.04/.22*	
Moderate index			−.06/.23*	
Yale Physical Activity Survey	25 adults, mean age 69.5 yr	2.5 days	kcal, r = .14 Activity summary, r = .37*	Dipietro, Casperson, Ostfeld, & Nadel, 1993
7-Day Recall	44 young adults	7 days	Counts, r = .35*	Dishman et al., 1992

[a]Correlation coefficient in parentheses was calculated when the Caltrac score was adjusted by adding an estimate of energy expenditure for the brief time the Caltrac was not worn by the subject.
*Statistically significant, $p < .05$.

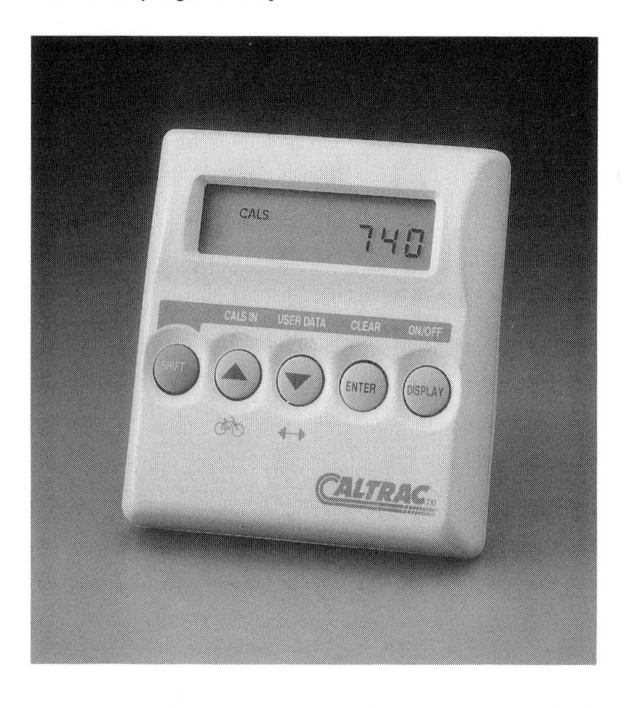

Figure 7.7 The Caltrac portable vertical accelerometer. Marketed by Muscle Dynamics Fitness Network, Inc., Torrence, CA, U.S.A.

females). The average accelerometer score was correlated with observations of two observers and activity ratings by two teachers. Similarly, Ayen and Montoye (1988) mounted three Caltracs at right angles to simulate a triaxial accelerometer on 30 young adults while they engaged in the following activities: treadmill walking at 3 mph and 0% grade, treadmill running at 6 mph and 0% grade, half-knee bends, bench stepping, simulated floor mopping, and a squat thrust calisthenic exercise. $\dot{V}O_2$ was measured during the activities.

When only walking and running were included, the correlation coefficient between the vertically mounted Caltrac and $\dot{V}O_2$ in liters per minute was .97, with a standard error of estimate (SEE) of 0.19 L · min^{-1}. The problem occurs when the other activities are added—the coefficient then drops to .65 (SEE = 0.55 L · min^{-1}). The outputs of the three Caltracs were combined by means of a multiple linear regression equation and by calculating a vector. The vector approach gave results only slightly better than the single-plane vertical accelerometer. The best approach was the multiple regression using the three Caltrac outputs as the independent variables. This raised the correlation coefficient relating the regression score with $\dot{V}O_2$ to .78 and reduced the SEE by 20%.

Since these studies were done, a triaxial accelerometer has been developed by Meijer, Westerterp, Koper, and ten Hoor in Maastricht, The Netherlands (1989). The outputs of the accelerometers were recorded on a tape recorder also carried by the subject. Sixteen adults (aged 20-24) sat, stood, walked, and ran on a treadmill at three speeds while $\dot{V}O_2$ was measured and accelerometer outputs recorded. The relationship of the triaxial accelerometer with the energy expenditure in joules · min^{-1} · kg^{-1} body weight was good (SEE = 79.1 J · min^{-1} · kg^{-1}). In a field study, four adults, mean age 22 years, wore the accelerometer for a week. Although the accelerometer-estimated energy expenditure exceeded the diary recorded intake by about 30%, the correlation coefficient relating the two was .99.

Two years later this group reported on an improved version of their triaxial accelerometer, with a new data acquisition unit eliminating the need for a tape recorder (Meijer, Westerterp, Verhoeven, Koper, & ten Hoor, 1991). Both the sensor and data storage unit can be worn on a belt. Bench tests over a week indicated that the accelerometer responded accurately to changes in frequencies and amplitudes, but there was a decrease in sensitivity over the week of about

12%. This was thought to be due to properties of the piezoelectric sensors and occurred when the unit was not used. With reuse, the sensitivity increased again. This physical characteristic of piezoelectric sensors may well occur with other accelerometers such as the Caltrac but has not been studied. With four subjects walking at three speeds on the treadmill and wearing two accelerometers, interinstrument variability and test-retest reliability were investigated. The mean coefficient of variation over 22 observations was less than 5%. Test-retest reproducibility resulted in a mean difference of 20% at 3 km/hr to less than 10% at 7 km/hr. The coefficient of correlation was .98. Corrections to the individual accelerometers based on the bench test did not improve the performance of the accelerometers significantly.

This group next conducted a field study that probably represents the best validation experiment done to date (Meijer, Westerterp, Wouters, & ten Hoor, 1990). Energy expenditure during 1 week was measured by the doubly labeled water technique. The triaxial accelerometer developed in the researchers' laboratory was worn by the subjects, and their basal metabolic rate was measured in a metabolic chamber. In many but not all cases, the accelerometer was calibrated by the bench test before and after the field trial. Each subject participated in a sitting and walking test while wearing an accelerometer, and those data were used to correct the estimates of energy expenditure from the accelerometers in the field. The correlation coefficient between the corrected accelerometer readings and doubly labeled water energy expenditure minus the sleeping metabolic rate was .88 (SEE = 706 $kJ \cdot day^{-1}$). When the doubly labeled water estimate was converted to METs, this coefficient was .77 (SEE = 942 $kJ \cdot day^{-1}$). Both coefficients were statistically significant at $p < .001$. Finally, the introduction of a training program in order to run a half marathon increased the weekly energy expenditure (doubly labeled water) in seven sedentary adult males and females (Meijer et al., 1991). An accelerometer was also worn during the periods of energy expenditure determinations. The increase in energy expenditure minus sleeping metabolic rate was significantly correlated with the increase in accelerometer counts (r = .78).

This triaxial accelerometer was recently tested during sedentary activities and walking (Bouten, Westerterp, Verduin, & Janssen, 1993). The estimation of energy expenditure during sedentary activities was best predicted by the three-dimensional output (r = .82; $S_{y \cdot x} = 0.22\ J \cdot s^{-1} \cdot kg^{-1}$; $p < .001$). The estimation of energy expenditure during walking was best predicted by accelerometer output in the anterior-posterior direction (r = .96; $S_{y \cdot x} = 0.53\ J \cdot s^{-1} \cdot kg^{-1}$; $p < .001$). If all types of activities are combined, the three-dimensional output predicts energy expenditure most accurately as long as walking is minimal (r = .95; $S_{y \cdot x} = 0.70\ J \cdot s^{-1} \cdot kg^{-1}$).

Because it appeared that a triaxial portable accelerometer may classify people on the basis of habitual physical activity better than a single-plane instrument, the company that marketed the original Caltrac has now developed and is marketing a triaxial accelerometer. This instrument, called the *Tri Trac-R3D,* will record activity data in 1- to 15-min time windows and will store data for up to 31 days in solid state memory. The output can be obtained as X, Y, and Z axis plots as well as the triaxial vector plot as a function of time. Energy expenditure in kilocalories is also an output option. The dimensions of the Tri Trac-R3D are 11.1 by 6.7 by 3.2 cm, and it weighs approximately 8 oz. It is powered by a 9-volt battery that lasts an average of 83 days. At present, the device sells for $500 (for one to nine units; less when more units are purchased). The computer interface and software for displaying or printing the data (only one set is needed) sell for $500. The Tri Trac-R3D and computer package are available from Hemokinetics, Professional Products Division, 5930 Seminole Center Ct., Madison, WI, USA 53711.

SUMMARY

Pedometers are inexpensive, simple movement counters that can estimate habitual physical activity over a relatively long period without interfering with or modifying subjects' normal lifestyles (Stunkard, 1960). The measurement principle is based on the number of steps taken during locomotion. However, there are serious problems with reliability and validity. The available pedometers do not have equal sensitivities; hence, some show high deviations from the actual step rate, even among those of the same type. We make these recommendations with regard to pedometers:

- Each pedometer should be calibrated by adjusting the tension of the spring and thereafter checking the pedometer score against the actual step rate at different walking and running speeds.
- Pedometers offer a good estimate of physical activity if most body movements coincide with vertical displacements of the whole body center of gravity, as happens in walking, jumping, running, and stepping. But where activities that do not cause vertical displacements of the body (including cycling, skating, and rowing) are an important part of daily energy expenditure, pedometer scores underestimate physical activity.

- Other problems arise when pedometers are used to estimate energy expenditure: Pedometers total the number of vertical displacements of the body and do not distinguish, for instance, between type of steps caused by a short period of high-intensity running and a long period of low-intensity walking. Because activities of high-intensity require more energy and are more important for physical fitness and health, in some situations it may be judicious to change the sensitivity so that it measures only activities of a relatively high intensity (for example, running with a speed of 6 km/hr). Moreover, at the same time, the adjustment prevents registration of passive movements like driving in a car over a bumpy road and other vibrational artifacts that are not caused by physical activities.
- The Large-Scale Integrated Motor Activity Monitor (LSI), although more expensive than the pedometer, is better standardized and useful in more activities.

There is a sound theoretical basis for attempting to estimate physical activity or energy expenditure using a portable accelerometer. However, the instrument must be waterproofed to register swimming movements, and during cycling it must be worn on a lower limb, not on the trunk. Also, it will not reflect the added energy expenditure when weight is lifted or carried. If habitual physical activity is to be estimated and day-to-day or seasonal variation is expected in the subject's activity, then various days or seasons must be sampled. Pedometers and mechanical accelerometers (converted watches, for example) are difficult to standardize. It would appear to be better to use the piezoelectric principle, but there material fatigue sometimes causes error. An accelerometer appears to be well suited to reflect an increase in walking or running.

At this writing, the only portable, single-plane accelerometer designed to estimate physical activity or energy expenditure and available commercially is the Caltrac. The interinstrument variability of this instrument is low and validity is good in walking or running under controlled laboratory conditions. However, in the field, if kilojoules or kilocalories of energy expenditure in usual activity in a particular season are to be estimated, at least 3 days, including 1 weekend day, should be averaged. If average daily METs is wanted, 6 days of recordings, including a weekend day, should be averaged. The validity of the Caltrac has been difficult to assess in the field because of the lack of an acceptable criterion.

The evidence to date suggests that a triaxial accelerometer will provide a slightly better estimate of usual physical activity or energy expenditure than a single-plane accelerometer. More research in the development and use of accelerometers in this context is needed. It should be borne in mind, however, that if an instrument is to be used in epidemiologic studies, its cost has to be acceptable. With advances in technical development, the cost of a triaxial accelerometer may decrease, making the instrument useful in many epidemiologic studies.

REFERENCES

Amar, J. (1920). *The human motor.* New York: E.P. Dutton.

Anderson, K.L., Masironi, R., Rutenfranz, J., & Seliger, V. (1978). *Habitual physical activity and health,* Copenhagen: World Health Organization Regional Publications. (European Series No. 6)

Avons, P., Garthwaite, P., Davies, H.L., Murgatroyd, P.R., & James, W.P.T. (1987). Approaches to estimating physical activity in the community: Calorimetric validation of actometers and heart rate monitoring. *European Journal of Clinical Nutrition,* **42,** 185-196.

Ayen, T.G., & Montoye, H.J. (1988). Estimation of energy expenditure with a stimulated three-dimensional accelerometer. *Journal of Ambulatory Monitoring,* **1,** 293-301.

Ballor, D.L., Burke, L.M., Knudson, D.V., Olson, J.R., & Montoye, H.J. (1989). Comparison of three methods of estimating EE: Caltrac, heart rate, and video analysis. *Research Quarterly for Exercise and Sport,* **60,** 362-368.

Balogun, J.A., Fanna, N.T., Fay, E., Rossman, K., & Pozyc, L. (1986). Energy cost determinations using a portable accelerometer. *Physical Therapy,* **66,** 1102-1107.

Balogun, J.A., Martin, D.H., & Clendenin, M.A. (1989). Calorimetric validation of the Caltrac accelerometer during level walking. *Physical Therapy,* **69,** 501-509.

Barber, C., Evans, D., Fentem, P.H., & Wilson, M.F. (1973). A simple load transducer suitable for long-term recording of activity patterns in human subjects. *Journal of Physiology, (London),* **231,** 94p-95p.

Bassey, E.J., Blecher, A., Fentem, P.H., & Patrick, J.M. (1982). Daily physical activity monitored before, during and after a walking-programme in middle-aged subjects. In F.D. Stott, E.B. Raftery, D.L. Clement, & S.L. Wright (Eds.), *Proceedings of the 4th International Symposium on Ambulatory Monitoring, and the Second Gent Workshop on Blood Pressure Variability, 1981* (pp. 394-400). London: Academic Press.

Bassey, E.J., Bryant, J.C., Fentem, P.H., MacDonald, I.A., & Patrick, J.M. (1980). Customary physical activity in elderly men and women using long-term ambulatory monitoring of ECG and footfall. In F.D. Stott, E.B. Raftery, & L. Goulding (Eds.), *Proceedings of the 3rd International Symposium on Ambulatory Monitoring* (pp. 425-432). London: Academic Press.

Bassey, E.J., Dallosso, H.M., Fentem, P.H., Irving, J.M., & Patrick, J.M. (1987). Validation of a simple mechanical accelerometer (pedometer) for the estimation of walking activity. *European Journal of Applied Physiology,* **56,** 323-330.

Bassey, E.J., Fentem, P.H., Fitton, D.L., MacDonald, I.C., Patrick, J.M., & Scrivven, P.M. (1978). Analysis of customary physical activity using long-term monitoring of ECG and footfall. In F.D. Stott, E.B. Raftery, P. Sleight, & L. Goulding (Eds.), *Proceedings of the 2nd International Symposium on Ambulatory Monitoring* (pp. 207-219). London: Academic Press.

Bell, R.Q. (1968). Adaptation of small wrist watches for mechanical recording of activity in infants and children. *Journal of Experimental Child Psychology,* **6,** 302-305.

Bell, R.Q., Weller, G.M., & Waldrop, M.F. (1971). Newborn and preschoolers: Organization of behavior and relations between periods. *Monograph of the Society for Research in Child Development,* **36,** 1-145.

Benedict, F.G., & Murschhauser, H. (1915). *Energy transformations during horizontal walking.* (Publication No. 231). Washington, DC: Carnegie Institute.

Berkowitz, R.I., Agras, W.S., Korner, A.F., Kraemer, H.C., & Zeanah, C.H. (1985). Physical activity and adiposity: A longitudinal study from birth to childhood. *Journal of Pediatrics,* **106,** 734-738.

Bhattacharya, A., McCutcheon, E.P., Shvartz, E., & Greenleaf, J.E. (1980). Body acceleration distribution and O_2 uptake in humans during running and jumping. *Journal of Applied Physiology,* **49,** 881-887.

Bouten, C.V., Westerterp, K.R., Verduin, M., & Janssen, J.D. *Assessment of energy during sedentary activities by triaxial accelerometry.* Manuscript submitted for publication.

Broskoski, M.B., Pivarnik, J.M., & Morrow, J.R., Jr. (1991). Caltrac validity for estimating caloric expenditure in children [abstract]. *Medicine and Science in Sports and Exercise,* **23,** S60.

Brouha, L. (1960). *Physiology in industry.* Elmsford, NY: Pergamon.

Brouha, L., & Smith, P.E., Jr. (1958). Energy expenditure of motions [abstract]. *Federation Proceedings of the American Society for Experimental Biology,* **17,** 20.

Buss, D.M., Block, J.H., & Block, J. (1980). Preschool activity level: Personality correlates and development implications. *Child Development,* **51,** 401-408.

Cauley, J.A., LaPorte, R.E., Black-Sandler, R., Orchard, T.J., Slemenda, C.W., & Petrini, C.W. (1986). The relationship of physical activity to high density lipoprotein cholesterol in post-menopausal women. *Journal of Chronic Diseases,* **39,** 687-697.

Cauley, J.A., LaPorte, R., Sandler, R.B., Bayles, C., Petrini, A., & Slemenda, C. (1984). Physical activity and HDL-C subfractions in post-menopausal women [abstract]. *Medicine and Science in Sports and Exercise,* **16,** 176.

Cauley, J.A., LaPorte, R.E., Sandler, R.B., Schramm, M.M., & Kriska, A.M. (1987). Comparison of methods to measure physical activity in post-menopausal women. *American Journal of Clinical Nutrition,* **45,** 14-22.

Cavagna, G.A., & Margaria, R. (1966). Mechanics of walking. *Journal of Applied Physiology,* **21,** 271-278.

Cavagna, G., Saibene, F., & Margaria, R. (1961). A three-dimensional accelerometer for analyzing body movements. *Journal of Applied Physiology,* **16,** 191.

Cavagna, G.A., Saibene, F.P., & Margaria, R. (1963). External work in walking. *Journal of Applied Physiology,* **18,** 1-9.

Cavagna, G.A., Saibene, F.P., & Margaria, R. (1964). Mechanical work in running. *Journal of Applied Physiology,* **19,** 249-256.

Chirico, A.-M., & Stunkard, A.J. (1960). Physical activity and human obesity. *New England Journal of Medicine,* **263,** 935-940.

Colburn, T.R., Smith, B.M., Guarini, J.J., & Simmons, N.W. (1976). An ambulatory activity monitor with solid state memory. *Instrument Society of America, Transactions,* **BM 76322,** 117-122.

Cook, T.C., Washburn, R.A., LaPorte, R.E., & Traven, N.D. (1985). Physical activity and body mass index as predictors of high density lipoprotein cholesterol over a year period in an active population of postal carriers [abstract]. *Medicine and Science in Sports and Exercise,* **17,** 220.

Cotes, J.E., & Meade, F. (1960). The energy expenditure and mechanical energy demand in walking. *Ergonomics,* **3,** 97-119.

Dawson, W.W. (1959). An electronic tambour: The piezoelectric crystal. *American Journal of Psychology,* **72,** 279-282.

Dewhurst, D.J. (1977). Characterization of human limb movements by accelerometry. *Medicine, Biology, Engineering and Computers,* **15,** 462-466.

Dion, J.L., Fouillot, J.P., & Leblanc, A. (1982). Ambulatory monitoring of walking using a thin capacitive force transducer. In F.D. Stott, E.B. Raftery, D.L. Clement, & S.L. Wright (Eds.), *Proceedings of the 4th International Symposium on Ambulatory Monitoring, and the Second Gent Workshop on Blood Pressure Variability, 1981* (pp. 420-425). London: Academic Press.

Dipietro, L., Caspersen, C.J., Ostfeld, A.M., & Nadel, E.R. (1993). A survey for assessing physical activity among older adults. *Medicine and Science in Sports and Exercise,* **25,** 628-642.

Dishman, R.K., Darracott, C.R., & Lambert, L.T. (1992). Failure to generalize determinants of self-reported physical activity to a motion sensor. *Medicine and Science in Sports and Exercise,* **24,** 904-910.

Eaton, W.O. (1983). Measuring activity level with actometers: Reliability, validity, and arm length. *Child Development,* **54,** 720-726.

Eaton, W.O., & Keats, J.G. (1982). Peer presence, stress, and sex differences in the motor activity levels of preschoolers. *Developmental Psychology,* **18,** 534-540.

Fenster, J.R., Freedson, P.S., Washburn, R.A., & Ellison, R.C. (1989). The relationship between peak oxygen uptake and physical activity in 6- to 8-year-old children. *Pediatric Exercise Science,* **1,** 127-136.

Foster, F.G., & Kupfer, D.J. (1973). Psychomotor activity and serum CPK activity. *Archives of General Psychiatry,* **29,** 752-758.

Foster, F.G., & Kupfer, D.J. (1975a). Anorexia nervosa: Telemetric assessment of family interaction and hospital events. *Journal of Psychiatric Research,* **12,** 19-25.

Foster, F.G. & Kupfer, D.J. (1975b). Psychomotor activity as a correlate of depression and sleep in acutely disturbed psychiatric inpatients. *American Journal of Psychiatry,* **132,** 928-931.

Foster, F.G., Kupfer, D.J., Weiss, G., Lipponen, V., McPartland, R., & Delgado, J. (1972). Mobility recording and cycle research in neuropsychiatry. *Journal of Interdisciplinary Cycle Research,* **3,** 60-72.

Foster, F.G., McPartland, R.J., & Kupfer, D.J. (1977). Telemetric motor activity in children: A preliminary study. *Biotelemetry,* **4,** 1-8.

Foster, F.G., McPartland, R.J., & Kupfer, D.J. (1978a). Motion sensors in medicine. Part I: A report on reliability and validity. *Journal of Inter-American Medicine,* **3,** 4-8.

Foster, F.G., McPartland, R.J., & Kupfer, D.J. (1978b). Motion sensors in medicine. Part II: Application in psychiatry. *Journal of Inter-American Medicine,* **3,** 13-17.

Freedson, P.S. (1991). Electronic motion sensors and heart rate as measures of physical activity in children. *Journal of School Health,* **61,** 220-223.

Frost, J.D. (1978). Triaxial vector accelerometry: A method for quantifying tremor and ataxia. *IEEE Transactions,* **BME 25,** 17-27.

Gayle, R., Montoye, H.J., & Philpot, J. (1977). Accuracy of pedometers for measuring distance walked. *Research Quarterly,* **48,** 632-636.

Gibbs-Smith, C. (1978). *The inventions of Leonardo da Vinci.* London: Phaidon Press.

Gretebeck, R., & Montoye, H. (1990). A comparison of six physical activity questionnaires with Caltrac accelerometer recordings [abstract]. *Medicine and Science in Sports and Exercise,* **22,** S79.

Gretebeck, R., & Montoye, H. (1992). Variability of some objective measures of physical activity. *Medicine and Science in Sports and Exercise,* **24,** 1167-1172.

Gretebeck, R., Montoye, H., & Porter, W. (1991). Comparison of doubly labeled water method for measuring energy expenditure with Caltrac accelerometer recordings [abstract]. *Medicine and Science in Sports and Exercise,* **23,** S60.

Gretebeck, R.J., Montoye, H.J., & Porter, W. (1993). *Validation of a portable accelerometer for estimating energy expenditure using doubly labeled water.* Unpublished manuscript.

Gretebeck, R., Weisel, S., & Boileau, R. (1992). Assessment of energy expenditure in active older women using doubly labeled water and Caltrac readings [abstract]. *Medicine and Science in Sports and Exercise,* **24,** S68.

Groenewegen, J.D. (1979). *Adiposity, activity and energy intake.* PhD thesis, University of Leiden, Rotterdam, The Netherlands: Bronder.

Hagg, G.M. (1982). Ambulant monitoring of lifting and carrying work. In F.D. Stott, E.B. Raftery, D.L. Clement, & S.L. Wright (Eds.), *Proceedings of the 4th International Symposium on Ambulatory Monitoring, and the Second Gent Workshop on Blood Pressure Variability, 1981* (pp. 384-389). London: Academic Press.

Halverson, C.F., Jr., & Post-Gorden, J.G. (1984). Measurement of open-field activity in young children: A critical analysis. In A.R. Liss (Ed.), *Energy intake and activity* (pp. 185-203). New York: Alan R. Liss.

Halverson, C.F., Jr., & Waldrop, M.F. (1973). The relations of mechanically recorded activity level to varieties of preschool play behavior. *Child Development,* **44,** 678-681.

Haymes, E.M., & Byrnes, W.C. (1991). Comparison of walking and running energy cost using the Caltrac and indirect calorimetry [abstract]. *Medicine and Science in Sports and Exercise,* **23,** S60.

Herron, R.E., & Ramsden, R.W. (1967). A telepedometer for the remote measurement of human locomotor activity. *Psychophysiology,* **4,** 112-115.

Hunter, G.R., Montoye, H.J., Webster, J.G., Demment, R., Ji, L.L., & Ng, A. (1989). The validity of a portable accelerometer for estimating energy expenditure in bicycle riding. *Journal of Sports Medicine and Physical Fitness,* **29,** 218-222.

Irving, J.M., & Patrick, J.M. (1982). The use of mechanical pedometers in the measurement of physical activity. In F.D. Stott, E.B. Raftery, D.L. Clement, & S.L. Wright (Eds.), *Proceedings of the 4th International Symposium on Ambulatory Monitoring, and the Second Gent Workshop on Blood Pressure Variability, 1981* (pp. 369-376). London: Academic Press.

Jacobs, D.R., Jr., Ainsworth, B.E., Hartman, T.J., & Leon, A.S. (1993). A simultaneous evaluation of 10 commonly used physical activity questionnaires. *Medicine and Science in Sports and Exercise,* **25,** 81-91.

Janz, K.F. (1994). Validation of the CSA accelerometer for assessing children's physical activity. *Medicine and Science in Sports and Exercise,* **26,** 369-375.

Johnson, C.F. (1971). Hyperactivity and the machine: The actometer. *Child Development,* **42,** 2105-2110.

Josenhans, W.T. (1967). Muscle testing by accelerometry: A preliminary analysis. *Internationale Zeitschrift für Angewande Physiologie Einschliesslich Arbeitsphysiologie,* **24,** 121-128.

Kashiwazaki, H., Inaoka, T., Suzuki, T., & Kondo, Y. (1986). Correlations of pedometer readings with energy expenditure in workers during free-living daily activities. *European Journal of Applied Physiology,* **54,** 585-590.

Kastango, K.B., & Freedson, P.S. (1991). Validation of a kilocalorie conversion of Caltrac activity counts [abstract]. *Medicine and Science in Sports and Exercise,* **23,** S61.

Kemper, H.C.G. (1992). Physical development and childhood activity. In N.G. Norgan (Ed.), *Physical activity and health* (pp. 84-100). Cambridge: Cambridge University Press.

Kemper, H.C.G., & Verschuur, R. (1974). Relationship between biological age, habitual activity and morphological physiological characteristics of 12- and 13-year-old boys. *Acta Paediatrica Belgica,* **38** (Suppl.), 191-203.

Kemper, H.C.G., & Verschuur, R. (1977). Validity and reliability of pedometers in habitual activity research. *European Journal of Applied Physiology,* **37,** 71-82.

Kemper, H.C.G., Verschuur, R., Ras, K.G.A., Snel, J., Splinter, P.G., & Tavecchio, L.W.C. (1976). Effect of 5- versus 3-lessons-a-week physical education program upon the physical development of 12 and 13 year old school boys. *Journal of Sports Medicine and Physical Fitness,* **16,** 319-326.

Klesges, R.C., Coates, T.J., Moldenhauer-Klesges, L.M., Holzer, B., Gustafson, J., & Barnes, J. (1984). The FATS: An observational system for assessing physical activity in children and associated parental behavior. *Behavioral Assessment,* **6,** 333-345.

Klesges, L.M., & Klesges, R.C. (1987). The assessment of children's physical activity: A comparison of methods. *Medicine and Science in Sports and Exercise,* **19,** 511-517.

Klesges, R.C., Klesges, L.M., Swenson, A.M., & Pheley, A.F. (1985). A validation of two motion sensors in the prediction of child and adult physical activity levels. *American Journal of Epidemiology,* **122,** 400-410.

Kripke, D.F., Mullaney, D.J., Wyborne, V.G., & Messin, S. (1978). There's no basic rest activity cycle. In F.D. Stott, E.B. Raftery, P. Sleight, & L. Goulding (Eds.), *Proceedings of the 2nd Symposium on Ambulatory Monitoring* (pp. 105-118). London: Academic Press.

Kupfer, D.J., Detre, T.P., Foster, F.G., Tucker, G.J., & DelGado, J. (1972). The application of Delgado's telemetric mobility recorder for human studies. *Behavioral Biology,* **7,** 585-599.

Kupfer, D.J., & Foster, F.G. (1973). Sleep and activity in a psychotic depression. *Journal of Nervous and Mental Disorders,* **156,** 341-348.

Kupfer, D.J., Weiss, B.L., Foster, F.G., Detre, T.P., DelGado, J., & McPartland, R. (1974). Psychomotor activity in affective states. *Archives of General Psychiatry,* **30,** 765-768.

LaPorte, R.E., Black-Sandler, R., Cauley, J.A., Link, M., Bayles, C., & Marks, B. (1983). The assessment of physical activity in older women: Analysis of the interrelationship and reliability of activity monitoring, activity surveys, and caloric intake. *Journal of Gerontology,* **38,** 394-397.

LaPorte, R.E., Cauley, J.A., Kinsey, C.M., Corbett, W., Robertson, R., Black-Sandler, R., Kuller, L.H., & Falkel, J. (1982). The epidemiology of physical activity in children, college students, middle-aged men, menopausal females and monkeys. *Journal of Chronic Diseases,* **35,** 787-795.

LaPorte, R.E., Kuller, L.H., Kupfer, D.J., McPartland, R., Matthews, G., & Caspersen, C. (1979). An objective measure of physical activity for epidemiologic research. *American Journal of Epidemiology,* **109,** 158-168.

Lauter, S. (1926). Zur Genese der Fettsucht. *Deutsches Archiv für Klinica Medicine,* **150,** 315-365.

Loo, C., & Wenar, C. (1971). Activity level and motor inhibition: Their relationship to intelligence-test performance in normal children. *Child Development,* **42,** 967-971.

Losen, K., Ryan, W., & Doigener, F. (1991). The validity of the Caltrac personal activity computer. *Medicine and Science in Sports and Exercise,* **23,** S12.

Lovely, D.F., Berne, N., Solomonides, S.E., & Paul, J.P. (1982). A measurement system to study free range lower limb prosthetic loading. In F.D. Stott, E.B. Raftery, D.L. Clement, & S.L. Wright (Eds.), *Proceedings of the 4th International Symposium on Ambulatory Monitoring, and the Second Gent Workshop on Blood Pressure Variability, 1981* (pp. 414-419). London: Academic Press.

Maccoby, E.E., Dowley, E.M., Hagen, J.W., & Dagerman, R. (1965). Activity level and intellectual functioning in normal preschool children. *Child Development,* **36,** 761-770.

Mahoney, M., & Freedson, P.S. (1990). Assessment of physical activity from Caltrac and Baecke questionnaire techniques [abstract]. *Medicine and Science in Sports and Exercise,* **22,** S80.

Marsden, J.P., & Montgomery, S.R. (1972). A general survey of the walking habits of individuals. *Ergonomics,* **15,** 439-451.

Massey, P.S., Lieberman, A., & Batarseh, G. (1971). Measure of activity level in mentally retarded children and adolescents. *American Journal of Mental Deficiency,* **76,** 259-261.

McPartland, R.J., Foster, F.G., & Kupfer, D.J. (1976). A computer-compatible multi-channel event counting and digital recording system. *Behavioral Research Methods and Instrumentation,* **8,** 299-301.

McPartland, R.J., Kupfer, D.J., & Foster, F.G. (1976). The movement-activated recording monitor: A third-generation motor-activity monitoring system. *Behavioral Research Methods and Instrumentation,* **8,** 357-360.

Meijer, G.A.L., Janssen, G.M.E., Westerterp, K.R., Verhoeven, F., Saris, W.H.M., & ten Hoor, F. (1991). The effect of a 5-month endurance-training programme on physical activity: Evidence for a sex-difference in the metabolic response to exercise. *European Journal of Applied Physiology,* **62,** 11-17.

Meijer, G.A., Westerterp, K.R., Koper, H., & ten Hoor, F. (1989). Assessment of energy expenditure by recording heart rate and body acceleration. *Medicine and Science in Sports and Exercise,* **21,** 343-347.

Meijer, G.A.L., Westerterp, K.R., Verhoeven, F.M.H., Koper, H.B.M., & ten Hoor, F. (1991). Methods to assess physical activity with special reference to motion sensors and accelerometers. *IEEE Transactions of Biomedical Engineering,* **38,** 221-228.

Meijer, G.A.L., Westerterp, K.R., Wouters, L., & ten Hoor, F. (1990). Validity of the accelerometer in the field: A comparison with the doubly labeled water technique. In G.A.L. Meijer, *Physical activity, implications for human EE.* PhD thesis, University of Limburg, Meppel, The Netherlands: Krips Repro.

Miller, D.J., Freedson, P.S., & Kline, G.M. (1994). Comparison of activity levels using the Caltrac® accelerometer and five questionnaires. *Medicine and Science in Sports and Exercise,* **26,** 376-382.

Montoye, H.J., Servais, S.B., & Webster, J.G. (1986). Estimation of energy expenditure from a force platform and an accelerometer. In J. Watkins, T. Reilly, & L. Burwitz (Eds.), *Sport science* (pp. 375-380). London: E. & F.N. Spon.

Montoye, H.J., Washburn, R., Servais, S., Ertl, A., Webster, J.G., & Nagle, F.J. (1983). Estimation of energy expenditure by a portable accelerometer. *Medicine and Science in Sports and Exercise,* **15,** 403-407.

Morrell, E.M., & Keefe, F.J. (1988). The actometer: An evaluation of instrument applicability for chronic pain patients. *Pain,* **32,** 265-270.

Morris, J.R.W. (1973). Accelerometry—a technique for the measurement of human body movements. *Journal of Biomechanics,* **6,** 729-736.

Nolan, M., Danner, F., Dewalt, K., McFadden, M., & Kotchen, J.M. (1990). The measurement of physical activity in young children. *Research Quarterly for Exercise and Sports,* **61,** 146-153.

Pambianco, G., Wing, R.R., & Robertson, R. (1990). Accuracy and reliability of the Caltrac accelerometer for estimating energy expenditure. *Medicine and Science in Sports and Exercise,* **22,** 858-862.

Redmond, D.P., & Hegge, F.W. (1985). Observations on the design and specification of a wrist-worn human activity monitoring system. *Behavioral Research Methods, Instruments and Computers,* **17,** 659-669.

Reswick, J.B., Perry, J., & Antonelli, D. (1979). Preliminary evaluation of the vertical acceleration gait analyzer (VAGA). *Proceedings of the 6th Annual Symposium on the Control of Human Extremities.* Dubrovnik, Yugoslavia.

Riker, D.M. (1980). A portable multichannel recorder for vector accelerometry. *Proceedings of the 33rd ACEMB* (p. 185). Washington, DC.

Roby, J.J., Sallis, J.F., Kolody, B., Condon, S.A., & Goggin, K. (1992). Developing self-administered self-reports of children's physical activity. *Medicine and Science in Sports and Exercise,* **24,** S69.

Sallis, J.F., Buono, M.J., Roby, J.J., Carlson, D., McClelland, C., & Morris, J.A. (1990). Reliability and validity of the Caltrac accelerometer as a physical activity monitor for children [abstract]. *Medicine and Science in Sports and Exercise,* **21,** S112.

Sallis, J.F., Buono, M.J., Roby, J.J., Carlson, D., & Nelson, J.A. (1990). The Caltrac accelerometer as a physical activity monitor for school-age children. *Medicine and Science in Sports and Exercise,* **22,** 698-703.

Saris, W.H.M., & Binkhorst, R.A. (1977a). The use of pedometer and actometer in studying daily physical activity in man. Part I: Reliability of pedometer and actometer. *European Journal of Applied Physiology,* **37,** 219-228.

Saris, W.H.M., & Binkhorst, R.A. (1977b). The use of pedometer and actometer in studying daily physical activity in man. Part II: Validity of pedometer and actometer measuring the daily physical activity. *European Journal of Applied Physiology,* **37,** 229-237.

Saunders, J., Inman, N., & Eberhart, H.J. (1953). The major determinants in normal and pathological gait. *Journal of Bone Surgery, American,* **35A,** 543-558.

Schulman, J.L., & Reisman, J.M. (1959). An objective measure of hyperactivity. *American Journal of Mental Deficiency,* **64,** 455-456.

Schulman, J.L., Stevens, T.M., & Kupst, M.J. (1977). The Biomotometer: A new device for the measurement and remediation of hyperactivity. *Child Development,* **48,** 1152-1154.

Schutz, Y., Froideraux, F., & Jequier, E. (1988). Estimation of 25 h energy expenditure by a portable accelerometer [abstract]. *Proceedings, Nutrition Society,* **47A,** 23.

Servais, S.B., Webster, J.G., & Montoye, H.J. (1984). Estimating human energy expenditure using an accelerometer device. *Journal of Clinical Engineering,* **9,** 159-171.

Simons-Morton, B.G., & Huang, I.W. (1991). Heart rate monitor and Caltrac assessment of moderate-to-vigorous physical activity among pre-adolescent children [abstract]. *Medicine and Science in Sports and Exercise,* **23,** S60.

Stunkard, A.J. (196)). A method of studying physical activity in man. *American Journal of Clinical Nutrition,* **8,** 595-601.

Sunnegårdh, J., & Bratteby, L.-E. (1987). Maximal oxygen uptake: Anthropometry and physical activity in randomly selected sample of 8- and 13-year-old children in Sweden. *European Journal of Applied Physiology,* **56,** 266-272.

Sweetman, B.J., Edwards, G.S., & Anderson, J.A.D. (1978). A measurement of limb activity in a back pain study in industry. In F.D. Stott, E.B. Raftery, P. Sleight, & L. Goulding (Eds.), *Proceedings of the 2nd International Symposium in Ambulatory Monitoring* (pp. 231-238). London: Academic Press.

Tamura, T., Nakajima, K., Togawa, T., Wakabayashi, R., & Osana, M. (1985). Development of a new exercise test for children. *Medicine, Biology, Engineering and Computers,* **23,** 482-486.

Taylor, C.R., Heglund, N.C., McMahon, T.A., & Looney, T.R. (1980). Energetic cost of generating muscular force during running. *Journal of Experimental Biology,* **86,** 9-18.

Taylor, C.B., Kraemer, H., Bragg, D., Miles, L., Rule, B., Savin, M., & DeBusk, R. (1980). A new system for long-term recording and processing of heart rate and physical activity in outpatients [abstract]. *Circulation,* **6**(III), 288.

Taylor, C.B., Kraemer, H.C., Bragg, D.A., Miles, L.E., Rule, R., Savin, W.M., & DeBusk, R.F. (1982). A new system for long-term recording and processing of heart rate and physical activity in outpatients. *Computer and Biomedical Research,* **15,** 7-17.

Verschuur, R., & Kemper, H.C.G. (1980). Adjustment of pedometers to make them more valid in assessing running. *International Journal of Sports Medicine,* **1,** 95-97.

Verschuur, R., Kemper, H.C.G., & Storm-van Essen, L. (1987). Longitudinal changes in habitual physical activity and in physical fitness, body size and body composition of boys at age 12/13 and 16/17. In R. Verschuur, *Daily physical activity and health* (pp. 21-44). Haarlem, The Netherlands: B.V. Uitgeverij De Vrieseborch.

Washburn, R., Chin, M.K., & Montoye, H.J. (1980). Accuracy of pedometer in walking and running. *Research Quarterly for Exercise and Sports,* **51,** 695-702.

Washburn, R.A., Cook, T.C., & LaPorte, R.E. (1989). The objective assessment of physical activity in an occupationally active group. *Journal of Sports Medicine and Physical Fitness,* **29,** 279-284.

Washburn, R.A., Janney, C.A., & Fenster, J.R. (1990). The validity of objective physical activity monitoring in older individuals. *Research Quarterly for Exercise and Sport,* **61,** 114-117.

Washburn, R.A., & LaPorte, R.E. (1988). Assessment of walking behavior: Effect of speed and monitor position on two objective physical activity monitors. *Research Quarterly for Exercise and Sport,* **59,** 83-85.

Weiss, B.L., Kupfer, D.J., Foster, F.G., & DelGado, J. (1974a). Psychomotor activity in mania. *Archives of General Psychiatry,* **31,** 379-383.

Weiss, B.L., Kupfer, D.J., Foster, F.G., & DelGado, J. (1974b). Psychomotor activity, sleep and biogenic amine metabolites in depression. *Biological Psychiatry,* **9,** 45-54.

Wilkinson, P.W., Parkin, J.M., Pearlson, G., Strong, H., & Sykes, P. (1977). Energy intake and physical activity in obese children. *British Medical Journal,* **1,** 756.

Williams, E., Klesges, R.C., Hanson, C.L., & Eck, L.H. (1989). A prospective study of the reliability and convergent validity of three physical activity measures in a field research study. *Journal of Clinical Epidemiology,* **42,** 1161-1170.

Williams, K.R. (1985). The relationship between mechanical and physiological energy estimates. *Medicine and Science in Sports and Exercise,* **17,** 317-325.

Wong, T.C., Webster, J.G., Montoye, H.J., & Washburn, R. (1981). Portable accelerometer device for measuring human EE. *IEEE Transactions in Biomedical Engineering,* **BME-28,** 467-471.

Chapter 8

ESTIMATING ENERGY EXPENDITURE FROM PHYSIOLOGIC RESPONSE TO ACTIVITY

Because of the difficulties encountered in measuring $\dot{V}O_2$ in the field, there is interest in the simpler but less direct method—recording physiologic data associated with energy expenditure. Advancements in telemetry and other aspects of bioengineering have made such techniques more attractive.

From the beginning of their existence, humans must have observed that pulse rate and ventilation increase during strenuous activity. Systolic blood pressure, electromyographs, and body temperature are also roughly proportional to the intensity of exercise. All of these variables can be telemetered, or entered on portable recorders.

Body Temperature

Under laboratory conditions, core body temperature and energy expenditure are closely related (Berggren & Christensen, 1950). However, it may require 40 to 50 min at a given exercise intensity before the body temperature reaches a steady state, so energy expenditure cannot be estimated from body temperature during exercise of short duration. However, for a given $\dot{V}O_2$, body temperature is slightly higher in less-fit subjects and higher for leg exercise compared to arm exercise (Berggren & Christensen). Body temperature does not accurately reflect metabolic rate during strenuous exercise in hot and moist climates. Also, the measurement of core temperature in the field is inconvenient and not practical in most situations.

Blood Pressure

Systolic blood pressure increases about linearly with an increase in exercise intensity, but the increase is not the same in dynamic and static exercise. Portable blood pressure recorders are now available, but during strenuous exercise they are subject to considerable artifacts. Blood pressure is subject to emotional influences as well. In light of these facts, the estimation of energy expenditure in the field from blood pressure recording is not recommended at this time.

Ventilation

Over a wide range of ventilation, extraction of oxygen is fairly constant, so ventilation can be used to estimate

$\dot{V}O_2$ with an error of about ±10% (Edholm, 1966). Ventilation has been measured in freely moving subjects (Bloom, 1965), but the difficulties are about as great as measuring $\dot{V}O_2$ in that the subject must wear a face mask, or a mouthpiece and nose clip. Also, the relationship of ventilation and $\dot{V}O_2$ over the full range of exercise intensity is not linear, although it is true that for most daily activities, ventilation falls within the linear range (Durnin & Edwards, 1955). Individual $\dot{V}O_2$–ventilation curves are necessary for accurate results. Recently, a method has been described that does not require a mask or mouthpiece (Anderson & Frank, 1990). This battery-powered system uses two bands of fabric containing insulated wire; one encircles the thorax, the other the abdomen. The expansion and contraction of the bands reflect cross-sectional changes of the thorax and abdomen. Breath-by-breath ventilation is calculated from these changes. However, this equipment is in the experimental stage; no practical system of estimating energy expenditure in the field from ventilation is yet available.

Electromyography

Electrical discharge from contracting muscles can be recorded in the form of an electromyogram (EMG). Theoretically, the amplitude and frequency of these electrical discharges might be expected to be related to energy expenditure. EMGs from the thigh were recorded on a portable cassette recorder to estimate physical activity (Anastasiades & Johnston, 1990). However, differences in skin sensitivity among individuals and variations in the placement of electrodes may seriously limit the usefulness of this approach. In any case, the validity and reliability, even in the laboratory, have not been studied.

Heart Rate

Of the physiological variables, heart rate (HR) is the easiest to measure in the field. The relationship between HR and energy expenditure was shown as early as 1907, when Benedict reported that changes in pulse rate were correlated with changes in heat production in any one individual. He later suggested that pulse rate may provide a practical and satisfactory method for estimating total metabolism.

Murlin and Greer in 1914 confirmed Benedict's results. They measured respiratory metabolism and HR simultaneously in subjects who were resting and doing moderate work. Their results indicated that HR was a good index of oxygen consumption. Thus, when work can be carefully controlled (as, for example, on a treadmill or bicycle), $\dot{V}O_2$ and HR are closely related and the relationship is linear over much of the range when the measurements are taken on one individual (Figure 8.1; from Montoye, 1970). The linear relationship of HR with $\dot{V}O_2$ can be understood from the Fick equation: $\dot{V}O_2 = HR \cdot SV\ (a\text{-}\bar{v}O_2\ diff)$. Over a wide range of exercise, stroke volume and a-$\bar{v}O_2$ diff do not change greatly; consequently, the increase in HR reflects an increase in $\dot{V}O_2$. Some investigators have presented data showing that the relationship is not linear over the full range from rest to strenuous activity (Berg, 1971; Booyens & Hervey, 1960; Bradfield, Huntzicker, & Fruehan, 1969; Henderson & Prince, 1914; Malhotra, Sen Gupta, & Rai, 1963; Viteri, Tourin, Galicia, & Herrera, 1971; Warnold & Lenner, 1977). Most agree that during exercise HR is more consistent and there is a greater tendency toward linearity than when resting values are included. Unfortunately, many in highly developed societies spend most of their days in a quiet state or doing only very light work.

Instrumentation

Various devices are now available that allow for the continuous monitoring of HR with little interference with the subject's activity. In selecting a HR monitor, several characteristics should be considered. There are three common methods of monitoring the heart rate: by detecting the electrocardiogram (ECG) with chest electrodes, by detecting a weaker ECG signal in the hands (as used in some exercising equipment), and by detecting blood flow in the finger or earlobe.

As early as 1950, Müller and Reeh modified an ear oximeter to record pulse rate during activity. However, using chest electrodes, one can use the ECG signal to record HR with fewer artifacts. One of the first of these devices was the Holter (1961) recorder, in which the ECG and hence the HR was recorded on a tape carried by the subject. The ECG or HR can be telemetered, but that requires a receiver in the vicinity and someone to operate it. Other instruments have been devised for recording HR on a portable tape recorder (Fouillot et al., 1982; Goldberger, 1961; MacConnie, Gilliam, Geenan, & Pels, 1982; Mueller et al., 1986; Rutenfranz et al., 1977). Usually, the recording speed is low so that HR may be recorded for a relatively long time (24 hr, for example). The tape is then replayed at a higher speed. Often, a second channel on the tape recorder is reserved for an independent time marker so heart rates will be accurate. When compared to directly recorded ECG, the heart rates are generally faithfully recorded (for example, see Elfner, Buss, Heene, & Kraatz, 1988). Because the electrodes are usually connected to the

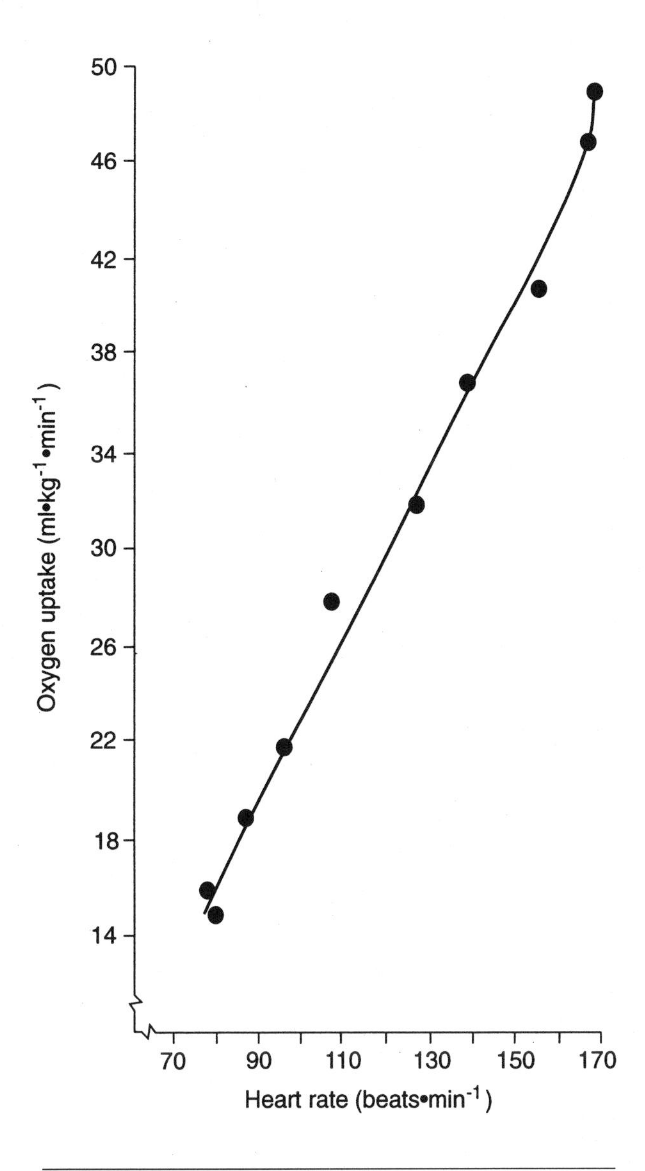

Figure 8.1 The relationship between a subject's oxygen uptake and heart rate while walking on a treadmill at increasing grades. From Montoye, H.J. (1970). Circulatory-respiratory fitness. In H.J. Montoye (Ed.), *An Introduction to Measurement in Physical Education*, Vol. 4, Chap. 3, p. 50. Indianapolis: Phi Epsilon Kappa Fraternity. Reproduced with permission.

recorder, problems of transmission are eliminated. The principal disadvantage of this system is that the unit is bulky and would be unsuitable for young children or during many vigorous activities. The tape recorder and playback systems are usually expensive.

Solid state HR recorders are also in general use, but because the ECG is not recorded, it is more difficult to detect errors in HR. These recorders are of three types: those that accumulate HR over time so that only an average rate is available, those that accumulate heartbeats in specific HR ranges (for example, eight levels), and those that record HR continuously.

Heartbeat accumulators. Among recorders that accumulate HR over time is the instrument described by Rowley, Glagov, and Stoner (1959) and Glagov, Rowley, Cramer, and Page (1970). It uses the R-wave of the ECG to move the hands of a modified wristwatch to accumulate heartbeats. It was developed in cooperation with engineers of Illinois Bell Telephone and Elgin National Watch companies. Compared with ECG recordings and manually determined HR, the error was about 1%.

A similar instrument was described by Friedman, Rosenman, and Brown (1963). Another HR integrator using an electrochemical approach was developed by Baker, Humphrey, and Wolff (1967); it is known as the Socially Accepted Monitoring Instrument (SAMI). This device records accumulated heartbeats using a gold electrode that becomes silver-plated in response to HR. Reversal of this process records the accumulated heartbeats. The main problem with accumulated heartbeats as an estimate of habitual activity is the fact that, for many people, the average HR over 24-hr is only a few beats per minute above the resting value. These low heart rates, as mentioned earlier, provide a poor estimate of energy expenditure. No estimates of time spent at higher heart rates are available.

Heart Rate Distribution Recorders. Heart rate distribution recorders constitute the second class of instruments. With these devices it is possible to accumulate heartbeats in a number of HR ranges. Data of this kind are obviously more valuable than simply a sum of all heartbeats. One of these instruments was developed by one of this book's authors and colleagues (Saris, Snel, & Binkhorst, 1977). The beat-to-beat intervals are recorded in seven registers ranging from 40 to 225 beats per minute, and the unit weighs only 220 g. When results were compared to a beat generator and taped ECGs during rest and vigorous exercise, the portable unit was very accurate. A similar device described by Mansourian, Masironi, Nicoud, and Steffan (1975) also appears to accurately accumulate heartbeats in eight registers. Data from these units must be retrieved by a readout unit.

Continuous Heart Rate Recorders. Recent years have seen many advances in the development of continuous HR recorders, the third type. Their performance has improved, and they are reduced in size and weight. Saris, Snel, Baecke, Van Waesberghe, and Binkhorst (1977) developed an instrument for monitoring minute-by-minute HR for up to 48 hr and storing this information in memory. Again, comparison of results with the pulse generator and tape-recorded heart rates in 14 subjects was excellent (mean error, .48% and .76%, respectively). The weight of the recorder is 195 g and measures 63 by 94 by 22 mm. A number of similar devices using the ECG signal to trigger the beat-by-beat HR have been developed and produced commercially. One of the most popular, the Sports Tester PE 3000, was studied under controlled conditions in which the ECG was simultaneously recorded (Treiber et al., 1989). In three experiments with 10 ten-year-olds, 23 children 4 to 6 years old, and 14 children 7 to 9 years old doing various exercises, the correlation coefficients between the ECG heart rates and the Sports Tester heart rates ranged from .94 to .99, with standard errors of measurement ranging from 1.1 to 3.7 beats per minute.

The most comprehensive lab testing of a number of HR monitors was done by Léger and Thivierge (1988). Eight men and two women served as subjects while the monitors were compared with ECG recordings during bicycle ergometer exercise up to 95% of estimated maximum HR. In the "excellent" category (r = .93 to .98; $S_{y.x}$ = 3.7% to 6.8%) were AMF Quantum XL (another name for the Sports Tester PE 3000), Exersentry, Pacer 2000H, and Monark 1. The "adequate" category (r = .84, $S_{y.x}$ = 11.7%) included only Seiko 1. In the third, "inadequate," category (r < .65; $S_{y.x}$ > 15%) were Accusplit 920, Tunturi Pulsemeter TPM 2000, Instapulse, Seiko 2, CIC Pulseminder 7719H, Sanyo Pulsemeter HRM-97E, and Monark 2.

The Sports Tester PE 3000 (AMF Quantum XL) was rated the best. The unit contains a small, lightweight transmitter (43 g, 137 by 30 by 12 mm) that attaches with two electrodes to the chest. A rubber chest band with electrodes can also be used. The receiver-recorder is a lightweight wristwatch (47 g; 51 by 45 by 15 mm) worn by the subject. An optional, immersible transmitter is available for swimmers. The unit can be programmed to record HR every 5 s for 80 min, every 15 s for 4 hr, or every 60 s for 16 hr. The latest version of the Sports Tester is even smaller (47 by 43 by 11 mm) and lighter (40 g), and the memory for 60-s recording has been extended to 34 hr and 40 min. In addition to instantaneous HR, the wristwatch receiver modes include time-of-day, a stopwatch function, elapsed time, and a HR alarm. The HR in memory during the recording period can be played back manually at various speeds. An interface and software are also available that, when connected to a computer, print out every HR or provide a graph of heart rates. The unit is manufactured by Polar Electro OY, Hukamaantie 18, SF-90440, Kempel, Finland, with distributors in various countries.

We have also had experience with the Sports Tester PE 3000, (AMF Quantum XL), which is also known under a third name, Uniq Heart Watch (Gretebeck, Montoye, Ballor, & Montoye, 1991). Thirty employed men wore the unit during their waking hours for 7 days. In general, the units operated very satisfactorily. For any day in the week, the percentage of heart rates missed varied from 1.9 to 2.4. There were a few circumstances during this and other studies when a satisfactory HR was not recorded, as shown in Table 8.1

Users of the device are not usually aware of how HR is determined, and this information is not included in the directions provided. Each displayed and stored HR represents a moving average. The following quotation illustrates how this is done.

> The pulse measuring algorithm calculates the first reading according to the first four pulse values. Thereafter, every other result is taken into account by averaging. When calculating a new mean value, the previous average result has the weight of 1/2; the latest result has the weight of 1/16. The averaging time is therefore dependent on the heart rate. At rest, when moment-to-moment heart rate changes frequently, the averaging time is as long as 15 seconds. During exercise, when heart rate is high, the averaging time is radically shortened so that the reading closely follows the moment-to-moment heart rate. For example, at a heart rate of 150 beats/min, the averaging time is only three to four seconds. (Karvonen, Chwalbinska-Moneta, & Saynajakangas, 1984, p. 68)

In monitoring HR, several precautions should be borne in mind. Both preparation of the skin and placement of the electrodes are important. The peak of the QRS complex can be maximized with respect to the rest of the ECG complex by judicious placement of electrodes. If the electrodes are placed over bone (sternum or ribs), muscle interference is minimized. Skin resistance can be reduced by being "roughed up" with

Table 8.1 Circumstances When Satisfactory Heart Rate Was Not Recorded Using the Polar Sports Tester PE 3000

Circumstance	Explanation
Driving auto (occurred in only one subject)	Drove with arm extended: lost signal
Flying private airplane	Electrical interference from radio equipment
Proximity of microwave oven or computer[a]	Electrical interference
Traffic signals[a]	Electrical interference
Operation of chain saw	Vibration
Mountain ski touring	Electrical interference from safety locator transceivers
Motorcycle riding	Lost signal

[a]The heart rate was not lost on every occasion, just occasionally.

abrasive material to remove the epithelium and then removing oil on the skin at electrode sites with alcohol. Abrasion of the skin increases the risk of irritation and infection, which must be minimized by careful selection of electrodes and gel. Excellent electrodes are available, containing gel that is placed in the electrode well. Double-sided adhesive disks are also available to hold the electrodes in place. A good discussion of the use of chest electrodes is contained in the report by Hanish, Neustein, Van Cott, and Sanders (1971).

Some of the other features of HR monitors, such as high and low heart rate alarm limits, are not necessary for assessing physical activity but may be useful if the monitor is to be used for other purposes (e.g., maintaining training intensity or avoiding overload during exercise).

Reproducibility of the Heart Rate

It is clear that convenient, portable equipment is available to dependably record and store HR during periods of 24 hr or more. However, if these data are to be used to estimate habitual physical activity, the day-to-day variation in human subjects in the field must be considered. This includes the question of how many days of recordings must be averaged for the values to reflect habitual heart rates.

When the same subject is tested several times under the same controlled conditions in an air-conditioned laboratory, the reproducibility of $\dot{V}O_2$ at various workloads is good (see Figure 8.2; from Montoye, 1970). This has been reported by others (Durnin & Namyslowski, 1958; Erickson, Simonson, Taylor, Alexander, & Keys, 1945). However, even under controlled conditions, there can be day-to-day variation in HR during exercise of the same workload (Figure 8.2). This has also been reported by others (Berg & Bjure, 1970; Christensen, Frey, Foenstelien, Aadland, & Refsum, 1983; Davies, Tuxworth, & Young, 1970; Maxfield, 1971; Montoye, Cunningham, Welch, & Epstein, 1970; Morgan & Pollock, 1978; Torún, 1984). But other investigations found little day-to-day variation in HR (Booyens & Hervey, 1960; Bradfield, 1979; Hellstrom & Holmgren, 1966; Kappagoda, Linden, & Newell, 1979; Purvis & Morgan, 1978; Rhyming, 1954; Shephard, 1969; Sime, Whipple, Berkson, MacIntyre, & Stamler, 1972; Warnold, Carlgren, & Krotkiewski, 1978; Washburn & Montoye, 1985). Reproducibility of HR during arm exercise appeared to be poorer than during leg exercise (Washburn & Montoye, 1985), probably because the arm exercise was less familiar to subjects.

Of interest is the repeatability of HR under field conditions. Thirty children aged 8 to 13 years wore HR meters (Uniq Heart Watch) for 2 days, an average of 10.5 hr per day (Sallis, Buono, Roby, Carlson, & Nelson, 1990). The mean of the lowest five heart rates for each day was subtracted from the day's mean HR for each child to calculate the activity HR. This method essentially eliminated the inherent difference in resting HR. The mean activity HR for the 30 subjects did not differ significantly from Day 1 to Day 2, but the correlation coefficient was only .10. Using the same equipment, heart rates were recorded on 110 children aged 3 to 5 years for 2 days, but with 3 to 6 months intervening. The correlation coefficient was .66 for the mean daily

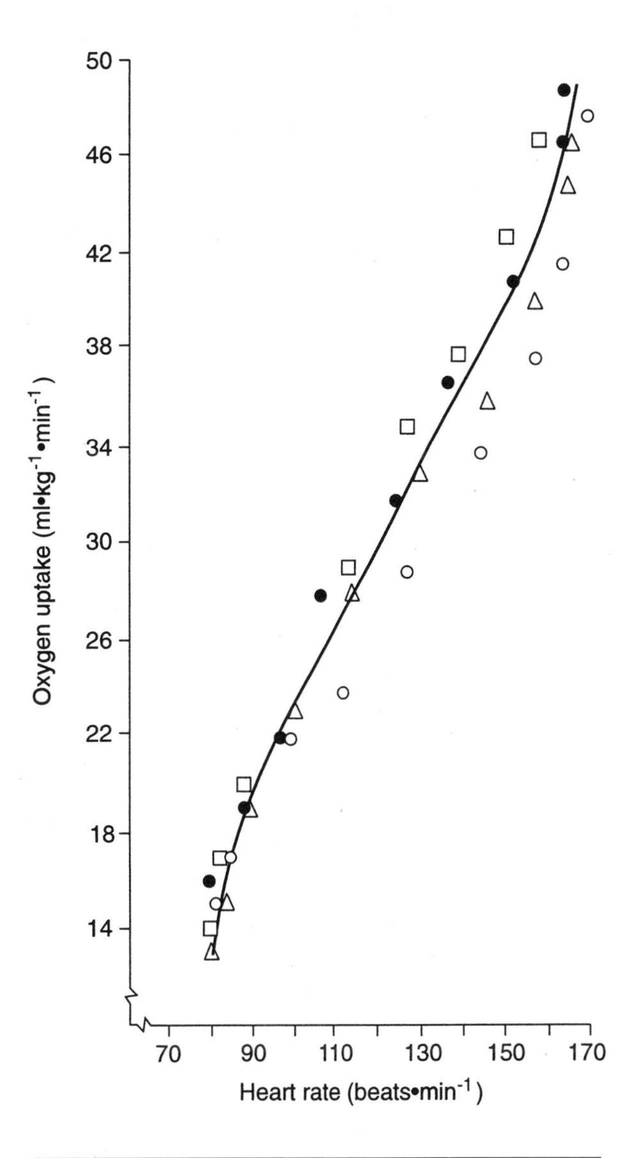

Figure 8.2 The relationships between oxygen uptake, heart rate, and treadmill grade (workload) in one subject tested four times, with at least a week between tests. •, First test; ❐, second test; Δ, third test; ❍, fourth test. From Montoye, H.J. (1970). Circulatory-respiratory fitness. In H.J. Montoye (Ed.), *An Introduction to Measurement in Physical Education*, Vol. 4, Chap. 3, p. 56. Indianapolis: Phi Epsilon Kappa Fraternity. Reproduced with permission.

HR (Durant et al., 1992). The coefficient was almost the same (.65) for the percentage of heart rates above 120. The authors estimate by the Spearman-Brown Prophecy formula that about 4 days of recording are required to achieve a reliability coefficient of .80. In these analyses, the inherent resting HR (r = .82) is not excluded, which would tend to raise the reported coefficients compared to correlating activity heart rates.

In a later study (Durant et al., 1993) a similar analysis was done with 131 older children, aged 5 to 7 years. With 3 to 6 months intervening and using 2 days of measurement, the intraclass correlation coefficient was .47 for mean daily HR and .67 for percentage of HR above 120. Again, using the Spearman-Brown Prophecy formula, it was estimated that about 4 days of measurement were needed to achieve a correlation coefficient of .80 for either mean HR or percentage of HR over 120. The results are similar to those reported by Gretebeck and Montoye (1992) on the reproducibility of HR recording in 30 employed men. Each subject wore a Uniq Heart Watch during the waking hours for 7 consecutive days. In addition to a mean daily HR, a HR index was calculated by dividing mean daily HR by baseline HR (mean of lowest 50 heart rates of the day). There were no significant differences in day-of-the-week mean HR or HR index. The average rank order correlation coefficients between 2 days were .62 and .44, respectively, for the mean HR and HR index. It was estimated by the Spearman-Brown Prophecy formula that an average of 5 days would produce reliabilities of .94 for the mean HR and .82 for the HR index.

Glagov et al. (1970) compared the average HR during 24 hr on 2 weekdays in 100 subjects. The intraindividual variance was significantly less than the interindividual variance. Richardson (1971) compared the HR of 21 males in two 24-hr periods. The means were almost identical for the 2 days (79.9 vs. 79.1), and the correlation coefficient was .69 ($p < .01$). Mueller and others (1986) studied day-to-day variability of the average HR in 64 sedentary middle-aged men. The HR recorder was worn on average 13 to 15 hr per day for 3 days (presumably weekdays) during each of 2 weeks. The Spearman rank order correlation for mean HR between any 2 days within the same week averaged .77 (Week 1) and .73 (Week 2). When the average of 3 days of Week 1 was compared with the average of the 3 days of Week 2, the Spearman rank order correlation coefficient was .85.

This group also used an index for each subject that consisted of the percentage of the total recording time in which the HR was above a specified level. This was thought to more accurately reflect the activity of the subject. For this index, the day-to-day rank order correlation coefficients were only .29 in both weeks. The correlation coefficient for the average of 3 days of Week 1 compared with the average of 3 days in Week 2 was .33.

Bassey, Bryant, Fentem, MacDonald, and Patrick (1980) also studied the day-to-day variability of a type of HR index in 57 men and women during working days. For most subjects HR was recorded for 15 to 16 hr a day. Subjects were brought into the laboratory, where the HR at which each walked at 4 km per hour was determined. An index was then developed consisting of the percentage of awake time spent at HR equal to or greater than that HR. When Day 1 was compared with Day 2, the correlation coefficient for the index was .61.

From these reports, it would seem that at least 4 days of HR recording are needed to reach a reliability coefficient of about .80 for the mean daily HR. Probably at least 6 days of recording would be needed to reach that level of reliability for a HR index (activity HR). If this is to reflect the heart rates over the year, the days of recording should be spread out over the year.

Heart Rate as an Estimate of Activity or Energy Expenditure in Adults

As we mentioned in the previous section, there is some day-to-day variation in HR at a given energy expenditure. To this must be added other sources of error. High ambient temperature and humidity or emotion may raise the HR with little effect on oxygen requirement of the work. Recently Hebestreit, Zehr, McKinty, Riddell, and Bar-Or (1993) quantified the influence of climatic heat stress on children's HR in order to reduce the error in estimating energy expenditure from HR. With HR at a temperature of 22 °C as standard, the following equation was developed for correcting the HR (eliminating the effect of heat stress) in order to estimate energy expenditure: $HR_{22\,°C} = HR_T\ (1.175 - [0.008 - T])$, where T is any given ambient temperature in °C, and HR_T is the measured HR at that temperature. This equation might be especially useful when energy expenditure is to be estimated from HR in a hot climate.

Training lowers the HR at which tasks of a given energy cost are performed. For example, active workers exercise at lower rates than sedentary men when the workload is equal (Taylor, 1967; Taylor & Parlin, 1966). Females have higher rates during exercise than males (Montoye, 1975). Fatigue (Booyens & Hervey, 1960; Lundgren, 1946) and state of hydration (Lundgren) affect the HR-$\dot{V}O_2$ relationship. Heart rates are higher for a given energy expenditure in anemic children (Gandra & Bradfield, 1971). Furthermore, certain kinds of activities, such as work with the arms only, will elicit higher heart rate than work done with the legs and arms, even though the oxygen cost is the same (Anderson, Liv,

Stamler, Van Horn, & Hoeksema, 1981; Collins, Cureton, Hill, & Ray, 1991; Durin & Namyslowski, 1958; Payne, Wheeler, & Salvosa, 1971; Vokac, Bell, Bautz-Holter, & Rodahl, 1975). Andrews (1971) has shown that the $\dot{V}O_2$-HR slopes were the same for arm and leg exercise but the intercepts were different. Static exercise increases HR above that expected on the basis of oxygen requirement (Hansen & Maggio, 1960; Maas, Kok, Westra, & Kemper, 1989).

Saris, Baecke, and Binkhorst (1982) showed that over 5 hr, changing the strenuousness of activities has an effect on the accuracy of the HR-to-energy-expenditure conversion, especially for quiet activities after moderate exercise: The energy expenditure is overestimated. This phenomenon may contribute to the overestimation of total energy expenditure compared to, for example, the doubly labeled water method, regardless of what $\dot{V}O_2$-HR regression equation is used.

Heart Rate-Oxygen Consumption Calibration Curves. Individual differences in HR response to exercise and differences in the inherent resting HR have rendered a general $\dot{V}O_2$-HR curve subject to large errors in estimating energy cost of activities (Andrews, 1971; Payne, Wheeler, & Salvosa, 1971). This has led to the common practice of bringing each subject into the laboratory to establish a $\dot{V}O_2$-HR curve and then using these curves to convert HR recorded in the field into $\dot{V}O_2$ values (Anderson et al., 1981; Andrews, 1971; Berg, 1971; Bradfield, 1971; Bradfield, Chan, Bradfield, & Payne, 1971; Bradfield, Paulos, & Grossman, 1971; Corbin & Pletcher, 1968; Gandra & Bradfield, 1971; Heywood & Latham, 1971; Payne, Wheeler, & Salvosa, 1971; Salvosa, Paine, & Wheeler, 1971; Verschuur & Kemper, 1985; Viteri, Tourin, Galicia, & Herrera, 1971). The $\dot{V}O_2$-HR calibration curve is more accurate if activities typical of the subject's daily activities are used to develop the calibration curve (Acheson, Campell, Edholm, Miller, & Stock, 1980; Kemper, Van Aalst, Leegwater, Maas, & Knibbe, 1990; Saris, Baecke, & Binkhorst, 1982; Warnold & Lenner, 1977).

As we have noted, there is some variation in the response of HR to different kinds of exercise at the same $\dot{V}O_2$. Multiple calibration curves for various activities could be developed for each individual, but time, expense, and inconvenience to the subjects render this approach impractical. One could also develop two calibration curves, one each for arm exercise and leg exercise, for each individual. Then, by using a diary for recording arm or leg exercise and applying the appropriate calibration curve, $\dot{V}O_2$ could be predicted. But this procedure requires more subject cooperation and has not been shown to be more effective than just one curve (Washburn & Montoye, 1986). Perhaps in the future a combination of HR recording and measurement of movement patterns with accelerometers may give better predictions of energy expenditure (this is discussed later in this chapter). In developing individual $\dot{V}O_2$-HR calibration curves over the entire range of energy expenditure, including resting values, it is clear that the $\dot{V}O_2$-HR relationship is curvilinear (Berg, 1971; Booyens & Hervey, 1960; Bradfield, Huntzicker, & Fruehan, 1969; Collins & Spurr, 1990; Henderson & Prince, 1914; Malhotra, Sen Gupta, & Rai, 1963; Saris, Baecke, & Binkhorst, 1982; Viteri et al., 1971; Warnold & Lenner, 1977).

The lack of a proportional increase in oxygen consumption with increases in HR as a result of a change in posture (lying, sitting, standing) is understandable. This change in posture would result in a decreased stroke volume with a compensatory increase in HR that would not be reflecting entirely a change in oxygen consumption. It is also likely that at low levels of oxygen consumption, HR is more susceptible to emotional influences that would also result in a disproportionate increase in HR for a constant oxygen consumption. Nevertheless, except for resting values during lying, sitting, and standing, the $\dot{V}O_2$-HR relationship is linear over most of the range (Van Dale, Schoffelen, ten Hoor, & Saris, 1989). Comparing various $\dot{V}O_2$-HR prediction models, Saris, Baecke, and Binkhorst (1982) concluded that using two linear $\dot{V}O_2$-HR regression lines and considering the HR while standing as a transition point is a practical solution. This approach was also used by Livingston et al. (1990) in comparing this so-called flex method with doubly labeled water.

Saris, Baecke, and Binkhorst (1982) measured energy expenditure ($\dot{V}O_2$) and HR for 5 hr in 13 adults. During 24 hr (including the 5-hr experiment), the subjects maintained an activity diary. The results of 10 different $\dot{V}O_2$-HR equations were correlated with the energy expenditure measured during the 5-hr period or estimated from the 24-hr diary. The results are shown in Figure 8.3. Energy expenditure (EE) is calculated from HR during the 5-hr and the total 24-hr period by these 10 different relationships for each individual.

1. Linear regression equation based on the three calibration points on the treadmill.

2. Linear regression equation based on the three calibration points on the bicycle ergometer.

3. Linear regression equation based on the calibration points of lying, sitting, and standing and the three points on the treadmill.

4. Linear regression equation based on the calibration points of lying, sitting, and standing and three points on the bicycle ergometer.

5. To find the transition point from one relationship between HR and EE to another at the change from quiet

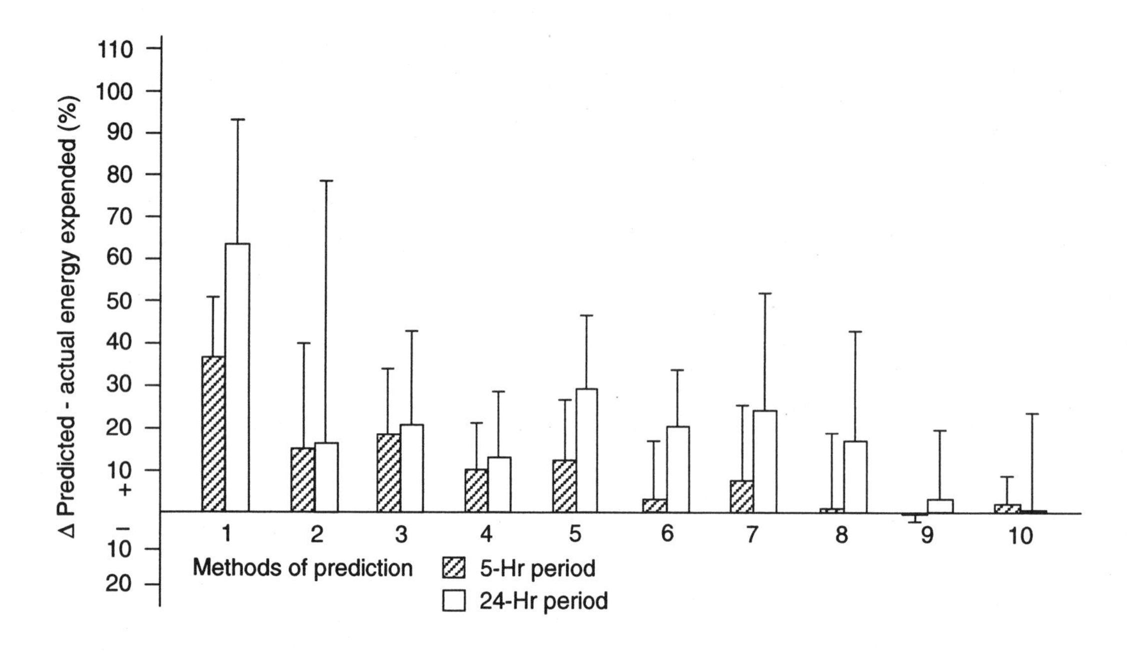

Figure 8.3 The mean and standard deviations (vertical bars) of differences between predicted and actual energy expenditure as a percentage of the actual values for the 5-hr and 24-hr periods for 10 methods of prediction. From Saris, W.H.M., Baecke, J., & Binkhorst, R.A. (1982). Validity of the assessment of energy expenditure from heart rate. In Saris, W.H.M., *Aerobic Power and Daily Physical Activity in Children*, pp. 100-117, PhD Thesis, University of Nymegen, Meppel: Kripps-Repro. Reproduced with permission.

activities to more dynamic activities, the results of the 5-hr period were plotted individually. From these plots the HR during standing proved to be a good indicator for this transition point, so the value for each subject was rounded off in line with the divisions of HR levels. (For example, a rate of 88 beats per minute during standing was rounded off to 90.) Below this transition point the linear regression equation based on the calibration points of lying, sitting, and standing was used. For the periods spent with a HR higher than this transition point, the linear regression equation of Calculation 1 was used.

6. HR lower than the HR transition point, see Calculation 5; HR higher than the HR transition point, see Calculation 2.

7. Because measuring HR and EE during lying, sitting, and standing is time-consuming, during the periods spent with a HR lower than the transition point, the EE value of the calibration point of standing was used instead of a regression equation. For HR higher than the HR transition point, see Calculation 1.

8. HR lower than the HR transition point, see Calculation 7; HR higher than the HR transition point, see Calculation 2.

9. EE was calculated from the mean HR during the experimental period and the individual regression line based on the 15 measurement points during the 5-hr period.

10. To reduce the number of measurement points, the mean HR was inserted in the individual linear regression line based on five representative measurement points during the 5-hr period: two quiet activities (standing and sitting—A and E, respectively), two moderate bicycle ergometer activities (C and N, respectively), and cycling free (F).

The use of net HR (work HR − resting HR) is a special case in this general approach. It is a simpler method of reducing the subject-to-subject variation, that is, correcting for resting metabolic rate, characteristic individual HR, and state of training. It is much like using one point on an individual calibration curve with the assumption that the slopes of the curves are the same for various individuals. Washburn and Montoye (1986) have demonstrated that this procedure is much simpler and is about as accurate as using a $\dot{V}O_2$-HR calibration curve on each subject. Goldsmith and Hale (1971) came to the same conclusion.

Another approach was suggested by Saris, Elvers, Van't Hof, and Binkhorst (1986). The level of 50% of $\dot{V}O_2max$ can be determined in the laboratory. The corresponding HR can then be used as a cutoff point for the more intense activity during a 24-hr HR profile, and energy expenditure above this level can be calculated. In this range (>50% energy expenditure) the estimation of energy expenditure from HR appears to be more valid than in the lower HR range.

Validity. Under controlled conditions, how accurately can one estimate EE from HR using individual $\dot{V}O_2$-HR calibration curves? When HR is used to estimate energy expenditure ($\dot{V}O_2$), the error is about 38% greater than the trial-to-trial measurement of $\dot{V}O_2$ when only one exercise is involved (Andrews, 1971). When different activities are involved, the error is about 70% greater. With 10 adult subjects performing various basketball activities for 30 min, the total kilocalories actually determined from $\dot{V}O_2$ were 187.4 ± 30.6, whereas the total kilocalories estimated from the heart rate were 218.0 ± 38.9, a statistically significant difference (Ballor, Burke, Knudson, Olson, & Montoye, 1989).

In several studies HR estimates of energy expenditure have been compared to the direct measurement using a whole-body calorimeter. In the investigation by Dauncy and James (1979), eight adult males individually occupied a room calorimeter for 27 hr while they carried out various activities, from resting to moderate energy expenditure. Several methods of developing individual calibration curves were studied. The best method involved calibrating within the chamber, but this is impractical in most situations. Otherwise, using a logarithmic curve was better than any of several straight line regressions in estimating energy expenditure. This method resulted in a mean difference in expenditure and HR-estimated expenditure for 24 hr of only about 6%, but the standard errors of prediction were quite large, about 20% to 25%.

Spurr et al. (1988) reported on the results of HR predictions of energy expenditure among 16 men and 6 women who each spent 22 hr in a calorimeter engaging in various activities. Before entering the calorimeter, each subject was calibrated to develop a $\dot{V}O_2$-HR curve in the usual way. Two linear regression equations were used to estimate energy expenditure. One was used when HR was at or below a critical value, this being the average of the highest resting HR and the lowest bicycle ergometer exercise HR + 10. The results averaged 86 ± 10 in males and 96 ± 6 in females. The second linear regression was developed for use when HR was above the critical value (the flex point). There was a mean difference of only 57 kcal for the total energy expenditure and 55 kcal in the expenditure during activity between the calorimeter value and that estimated from HR for the 22-hr period. The correlation coefficients were .87 (SEE = 216 kcal) and .80 (SEE = 216 kcal), respectively, for total and activity energy expenditure. In a second study by this group (Ceesay et al., 1989), 20 adults each spent 21.5 hr in the calorimeter. When the same methods as in the previous investigation were used, there was again an insignificant difference of 1.2% in the mean total energy expenditure, with a correlation coefficient of .943 (SEE = 109 kcal).

There have been some attempts to assess the validity of the HR method of estimating energy expenditure in adults in field studies. These are summarized in Table 8.2. Except for the doubly labeled water (DLW) method, the criteria in these studies leave much to be desired, and in many studies the HR was recorded for only 1 or 2 days, which likely contributed to poor estimates of validity. It should be pointed out that in the study by Livingston and others (1990) using the HR flex method (Table 8.2), a comparison of energy expenditure with DLW method resulted in a mean difference of only 2.0%, but the individual differences range from −22% to 52%. Nine of the 14 differences were within ±10%. In the other DLW validity study (Schultz, Westerterp, & Brück, 1989) listed in Table 8.2, a mean difference between the HR estimate of energy expenditure using a linear equation and the DLW method was 9.8% ± 20.3%. The individual differences range from −20.7% to 47.7%, with four out of nine values lying within 10%. Other procedures, such as second order regression of HR on $\dot{V}O_2$ or the use of two linear equations as in the flex method, did not yield better results, as was also found in the study with children by Emons, Groenenboom, Westerterp, and Saris (1992).

Heart Rate as an Estimate of Activity or Energy Expenditure in Children

Because of the difficulty experienced with diary and questionnaire methods of assessing habitual activity in children, the use of HR is especially appealing (Saris, 1986). Unfortunately, fewer data from studies using this technique are available in children than in adults. Just as in adults, it is clear that individual $\dot{V}O_2$-HR calibration curves or a heart rate index is essential if energy expenditure in the field is to be estimated. Also, as in adults, energy expenditure (kilojoules) is a poor indicator of physical activity because of the influence of body size; hence, if activity is to be estimated, energy expenditure must be expressed in METs or in terms of body size (kilojoules per kilogram of body weight, for example). A summary of the studies in which some type of validation was done in children is shown in Table 8.3.

Multiple Recording Systems to Estimate Energy Expenditure

Because of the limitations of any of the practical field methods of assessing energy expenditure and physical activity, there have been attempts to combine approaches in an effort to improve validity. For example, emotional effects can raise HR during periods of little or

Table 8.2 Validation of the Heart Rate (HR) Method of Estimating Energy Expenditure in Adults: Field Studies

Criterion	Population	HR method	Heart rate recording period	Correlation[a]	Mean difference	Reference
Doubly labeled water	6, ages 20-30	Individual calibration log plot	2 days	r = .73*	Insignificant	Schulz et al., 1989
Dietary intake (weighed)		$\dot{V}O_2$/HR		(SD = 11%) r = .80*		
Diary				r = .63*		
Doubly labeled water	14, ages 17-46	Individual calibration 2 Linear regression	2-4 days	SEE = 17.9%	Heart rate overestimated 2%, difference insignificant	Livingston et al., 1990
$\dot{V}O_2$ measured at work, 10-40 min, various times	20 male assembly workers, 14 female nurses	Linear regression	Unknown	r = .55* r insignificant	— —	Fordham, Goldsmith, Kaval, O'Brien, & Tan, 1978
Dietary intake (bomb calorimeter, 6-12 days)	12 male adults	Individual calibration log plot $\dot{V}O_2$/HR	35-315 days	SEE = 6.7%	HR overestimated 7.3%	Acheson et al., 1980
Dietary intake (weighed)	12 female adults	Individual calibration 2 Linear regression	4 days	r = .6*	HR overestimated 7%	Kalkwarf, Haas, Belko, Roach, & Roe, 1989
Diary (METs)	20 males, ages 19-60 yr	Individual linear regression	1 day	SEE = 39%	HR underestimated 19%	Washburn & Montoye, 1986
		Individual log $\dot{V}O_2$/HR		SEE = 24%	HR underestimated 22%	
		Mean HR minus resting HR		SEE = 23%	HR overestimated 11%	

[a]SD = standard deviation; SEE = standard error of estimate.
*Statistically significant, $p < .05$.

Table 8.3 Validation of the Heart Rate (HR) Method of Estimating Energy Expenditure in Children (Laboratory and Field Studies)

Criterion	Population	HR method	HR recording period	Correlation	Mean difference	Reference
Caltrac	35, ages 8-13 yr	HR minus resting HR	2 days	.49*	—	Sallis et al., 1990
Activity recall questionnaire			Day 1 Day 2	.25 .52*	— —	
Dietary intake (bomb calorimeter)	17, ages 4-5 yr, obese and nonobese	Individual calibration linear	4-7 days	—	HR overestimated kcal 3.8%	Griffiths & Payne, 1976
Dietary intake (recall)	801, mean age 8.2 yr	Individual calorie linear and standing HR	1 day	—	HR overestimated 1.9% (boys) HR overestimated 1.4% (girls)	Saris, Boeyen, Elvers, De Boo, & Binkhorst, 1982
Calorimeter	9 boys, mean age 9.3 yr 10 girls, mean age 8.1 yr	Individual calibration curve	24 hr	—	HR overestimated 10.4%	Emons et al., 1992
Doubly labeled water	5 boys, mean age 8.4 yr 5 girls, mean age 8.4 yr	Individual calibration curve	24 hr	—	HR overestimated 12.3%	Emons et al., 1992
Observation (physical education class)	36, ages 8-10 yr	HR during each activity	About 34 min	36 individual r's .29 to .90 mean r = .64	—	O'Hara, Baranowski, Simons-Morton, Wilson, & Parcel, 1989

(continued)

Table 8.3 *(continued)*

Criterion	Population	HR method	HR recording period	Correlation	Mean difference	Reference
Video analysis Caltrac	20, mean age 15.2 yr	Individual linear regression	Basketball class, 37 min	.89* .92*	HR overestimated 59% HR overestimated 20%	Ballor et al., 1989
PWC 170	171, ages 4-6 yr 54, ages 8-12 yr	Mean HR	1 day	Mean daily HR significantly lower in more fit	—	Saris, Binkhorst, Cramwinckel, Van der Veen-Hezemans, & Van Waesberghe, 1979
Estimated $\dot{V}O_2$max (treadmill)	800, ages 6, 8, and 10 yr	Individual calibration, linear regression	1 day	Mean daily HR significantly lower in more fit No significant difference in energy expenditure between fit and unfit	—	Saris, Noordeloos, et al., 1982; Saris, Elvers, Van't Hof, & Binkhorst, 1986
Motor ability test	32, ages 5-6 yr	Mean daily HR Mean HR minus resting HR	1 day	Mean lower in fit children Mean higher in fit children	—	Mimura, Hebestreit, & Bar-Or, 1991
$\dot{V}O_2$max	17, ages 9-11 yr	Time HR > lactate threshold**	.72*	—	—	Atomi, Iwaoka, Hatta, Miyashita, & Yamamoto, 1986
		Time HR > 60% max	.75*	—		

*Statistically significant, $p < .05$.
**The lactate threshold, sometimes called the anaerobic threshold, is the point in an exercise test of increasing intensity at which the blood level of lactic acid begins to rise exponentially. It is considered to occur at a higher workload in more fit subjects.

no physical activity. If one could exclude those episodes, perhaps HR would better reflect energy expenditure.

Bassey and colleagues at the university hospital and medical school in Nottingham, England, have devised a system of recording HR and footfall simultaneously (Bassey et al., 1978; Bassey et al., 1980; Patrick, Bassey, Irving, Blecher, & Fentem, 1986). A body-borne tape recorder was employed to record a time signal, HR from the ECG, and footfall from a pressure transducer under the heel. This was used in connection with a standard walking test so that the time during which HR equaled or exceeded HR during walking could be determined. This is a form of individual calibration. The system appears to be reliable and well received by the subjects. The records documented the increased walking performed by the subjects in response to a walking intervention program; although pedometers worn by the subjects also reflected the increased walking. The system also documented a decrease in physical activity following retirement. However, there was no attempt to validate the system using a suitable external criterion; apparently the objective was not to separate out emotional influences on HR. Also, only walking was of interest in these studies.

A portable multiple recording system was reported by Taylor and co-workers (Taylor et al., 1980, 1982) in which HR (from ECG) and the output from six mercury switches arranged on a cube were recorded and stored on a microcomputer. One of the purposes of the studies was to separate increased HR associated with and without physical activity, with the second assumed to be related to anxiety. The data points were collapsed into 5-min periods. Ten adult subjects were calibrated by treadmill exercise; they wore the equipment continuously for 24 hr, and during their waking hours they noted their activity and anxiety levels every 15 min. Simultaneous monitoring of ECG indicated that the HR recording system was accurate. However, there was poor agreement between instances of self-reported anxiety or lack of it and monitored anxiety (high HR, low activity) or lack of it (low HR and low activity, or high HR and high activity). Two of this group (Miles & Rule, 1981) reported on an expanded version of this Vitalog system (Vitalog PMS-8) in which ECG, respirations, movement, and body temperature can all be monitored. It is a solid-state system, weighing 12 oz (6 by 3.4 by 1.3 in.). Records of several individuals are discussed. Further validation studies of these systems are needed.

A four-channel portable tape recording system was described by Anastasiades and Johnston in 1990. One of the channels of the Medilog 4-24 cassette recorder (Oxford Medical Systems) was used for the ECG, a second for the electromyograph (EMG) from the thigh, and a third for marking events; the fourth contained a time module. A controlled situation was created to test the equipment. Both the ECG and EMG signals were accurately recorded. In field use with several hundred ambulatory subjects, the failure of the EMG system was less than 4%. The system was developed primarily to index variations in physical activity within one individual. Whether it is useful for evaluating physical activity between individuals is not known.

Recently Haskell, Yee, Evans, and Irby (1993) attached one motion sensor (a mercury switch) to the arm and one to the leg of 15 males aged 22 to 64 while they performed walking/running on the treadmill, arm cranking, cycling, and bench stepping in the laboratory. Heart rate was also measured, which, together with the output of the two motion sensors, was recorded on a Vitalog PMS-8 monitor (Vitalog Corp., Redwood City, CA, USA). Oxygen uptake was measured during the various activities. Including the output of the motion sensors in a regression equation improved the prediction of oxygen uptake compared to using only HR. Without using individual $\dot{V}O_2$-HR calibration curves, and pooling data of all subjects, the multiple correlation coefficient was .85 (SEE = 5.2 ml $O_2 \cdot kg^{-1} \cdot min^{-1}$). If the individual $\dot{V}O_2$-HR calibration curves were used, the average multiple correlation coefficient was .94 (SEE = 2.3 ml $O_2 \cdot kg^{-1} \cdot min^{-1}$).

Meijer, Westerterp, Koper, and ten Hoor (1989) observed four subjects during 1 week in which food intake was measured and heart rate and a three-dimensional accelerometer output were recorded. Figure 8.4 is an example of the simultaneous recording of HR and accelerometer output.

Multiple Methods of Measurement

Another approach to estimating energy expenditure is to combine different measurement methods for which the information overlaps only partly.

Kemper and his colleagues (Kemper, 1992) in the Amsterdam Growth and Health Longitudinal Study used three approaches to measure physical activity in teenagers: HR recording over 48 hr, pedometer scores over 2 weekdays and 1 weekend, and a standardized activity interview covering the previous 3 months, giving total time spent per week in three intensity levels based on METs.

Pearson correlation coefficients between these three activity instruments applied over 4 years of measurement in about 200 boys and girls varied between .16 and .20, each explaining only a very small part of the total variation. This illustrates that in this population the use of one of the three methods poorly estimates activity and a combination of measurements applied

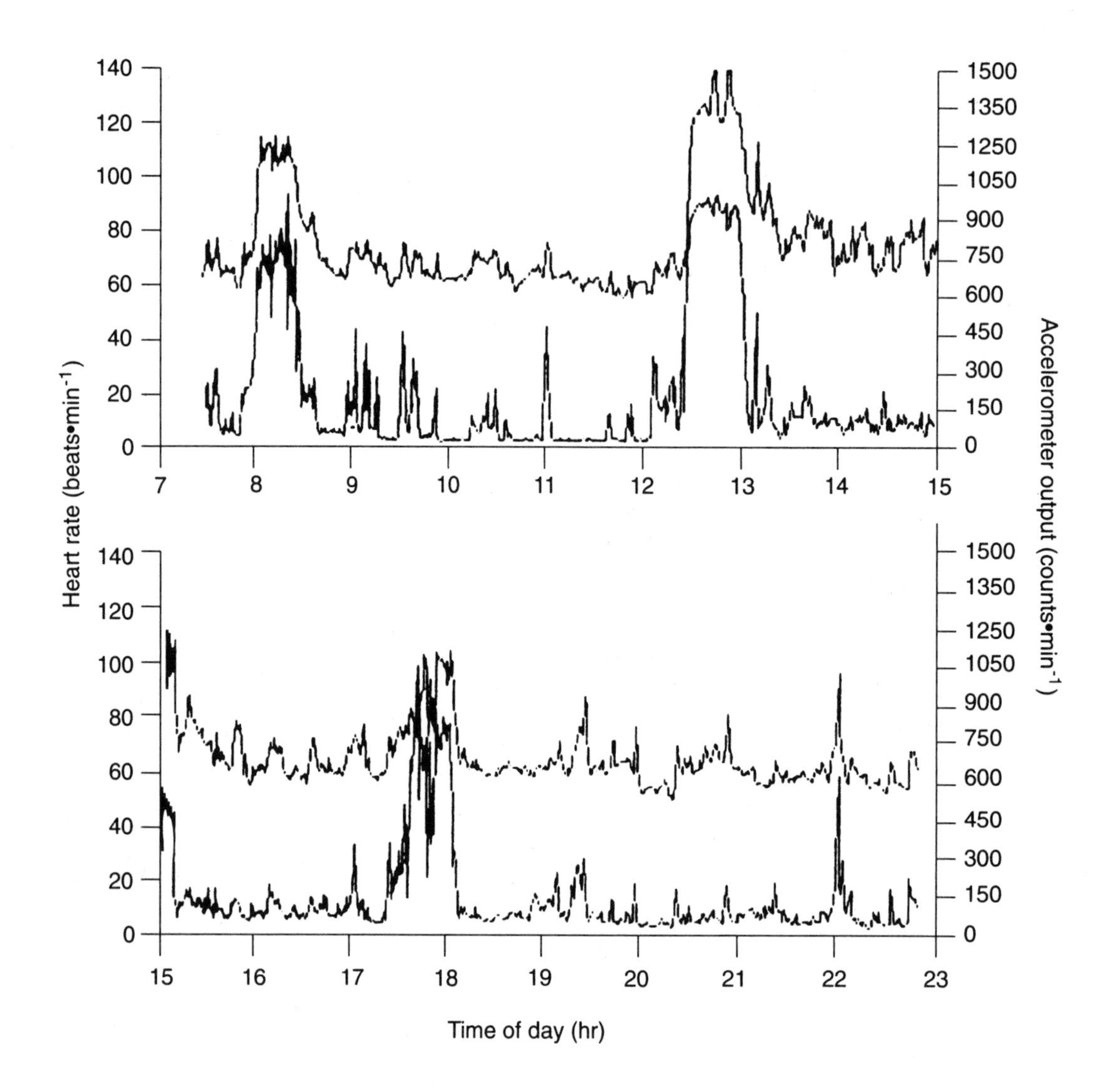

Figure 8.4 Daytime recording of heart rate (upper line, beats · min^{-1}) and accelerometer output (lower line, counts · min^{-1}) of one observation day. From Meijer, G.A.L., Westerterp, K.R., Koper, H., & ten Hoor, F. (1989). Assessment of energy expenditure by recording heart rate and body acceleration. *Medicine and Science in Sports and Exercise*, **21**, 343-347. Reproduced with permission of Williams and Wilkins Company.

over different periods is needed to obtain a more valid picture of the daily activity pattern.

To validate the activity measurements in this investigation, the $\dot{V}O_2max$ was compared between active and inactive subjects. All subjects were measured during 4 successive years. The 13- and 17-year-old teenagers were divided into active and inactive groups. Inactive were those who scored below and active those who scored above the median in two out of three methods of estimating physical activity during 3 out of 4 years.

Comparing the active and inactive teenagers over the years with respect to maximal aerobic power relative to body weight showed that the active girls and boys had significantly ($p > 0.05$) higher $\dot{V}O_2max$ adjusted for body weight between the age of 10 and 17 years than the inactive girls and boys.

SUMMARY

A number of physiological functions reflect the rate of energy expenditure, but HR is the most practical response to measure in the field. There are now dependable, self-contained, portable HR recorders available at reasonable cost. In populations where day-to-day variation has been studied, 4 or 5 days of recording (including a weekend day) appear to be necessary to obtain a HR index that is typical for an individual. In order to interpret HR as an index of physical activity or energy expenditure, it is imperative to employ individual $\dot{V}O_2$-HR calibration curves or to subtract the resting HR from recorded HR; the second method is simpler and probably almost as accurate.

Nevertheless, HR is affected by factors other than the intensity of the physical activity, the most significant being emotions, and thus leaves much to be desired as an index of physical activity or energy expenditure. It is probably most useful when other methods are not feasible—in young children, for example, or in combination with another method. If one is interested in the amount of moderate and intense physical activity, the estimation of energy expenditure above 50% of $\dot{V}O_2$max is a reasonable alternative. However, the determination of each individual's $\dot{V}O_2$max would be necessary.

The use of motion sensors in combination with HR recording seems to be a promising approach, but multiple systems are in the developmental stage, making it impossible at present to recommend practical applications for those systems. Clearly the cost would be prohibitive for many applications.

Another approach is to combine methods of estimating physical activity (questionnaire, HR recording, motion sensors, etc.), but the validity of this approach has yet to be established. The cost of course would be greater than if only one method were employed.

REFERENCES

Acheson, K.J., Campell, I.T., Edholm, O.G., Miller, D.S., & Stock, M.J. (1980). The measurement of daily energy expenditure—an evaluation of some techniques. *American Journal of Clinical Nutrition,* **33,** 1155-1164.

Anastasiades, P., & Johnston, D.W. (1990). A simple activity measure for use with ambulatory subjects. *Psychophysiology,* **27,** 87-93.

Anderson, D.E., & Frank, L.B. (1990). A microprocessor-based system for monitoring breathing patterns in ambulatory subjects. *Journal of Ambulatory Monitoring,* **3,** 11-20.

Anderson, R.M., Liv, K., Stamler, J., Van Horn, L., & Hoeksema, R. (1981). Assessment of individual physical activity level: Intra-individual versus inter-individual variability [abstract]. *Medicine and Science in Sports and Exercise,* **13,** 100.

Andrews, R.B. (1971). Net heart rates substitute for respiratory calorimetry. *American Journal of Clinical Nutrition,* **24,** 1139-1147.

Atomi, Y., Iwaoka, K., Hatta, H., Miyashita, M., & Yamamoto, Y. (1986). Daily physical activity levels in preadolescent boys related to $\dot{V}O_2$ max and lactate threshold. *European Journal of Applied Physiology,* **55,** 156-161.

Baker, J.A., Humphrey, J.E., & Wolff, H.S. (1967). Socially acceptable monitoring instrument (SAMI). *Journal of Physiology,* **188,** 4-5.

Ballor, D.L., Burke, L.M., Knudson, D.V., Olson, J.R., & Montoye, H.J. (1989). Comparison of three methods of estimating energy expenditure: Caltrac, heart rate, and video analysis. *Research Quarterly for Exercise and Sport,* **60,** 362-368.

Bassey, E.J., Bryant, J.C., Fentem, P.H., MacDonald, I.A., & Patrick, J.M. (1980). Customary physical activity in elderly men and women using long-term ambulatory monitoring of ECG and footfall. In F.D. Stott, E.B. Raftery, & L. Goulding (Eds.), *Proceedings of the 3rd International Symposium on Ambulatory Monitoring* (pp. 425-432). London: Academic Press.

Bassey, E.J., Fentem, P.H., Fitton, D.L., MacDonald, I.C., Patrick, J.M., & Scrivven, P.M. (1978). Analysis of customary physical activity using long-term monitoring of ECG and footfall. In F.D. Stott, E.B. Raftery, P. Sleight, & L. Goulding (Eds.), *Proceedings of the 2nd International Symposium on Ambulatory Monitoring* (pp. 207-219). London: Academic Press.

Benedict, F.G. (1907). *The influences of inanition on metabolism.* (Publ. No. 77). Washington, DC: Carnegie Institute.

Berg, K. (1971). Heart rate telemetry for evaluation of the energy expenditure of children with cerebral palsy. *American Journal of Clinical Nutrition,* **24,** 1431.

Berg, K., & Bjure, J. (1970). Methods for evaluation of the physical working capacity of school children with cerebral palsy. *Acta Paediatrica Scandinavica (Suppl.),* **204,** 15-26.

Berggren, G., & Christensen, E.H. (1950). Heart rate and body temperature as indices of metabolic rate during work. *Arbeitsphysiologie,* **14,** 255-260.

Bloom, W.L. (1965). A mechanical device for measuring human energy expenditure. *Metabolism: Clinical and Experimental,* **14,** 955-958.

Booyens, J., & Hervey, G.R. (1960). The pulse rate as a means of measuring metabolic rate in man. *Canadian Journal of Biochemical Physiology,* **38,** 1301.

Bradfield, R.B. (1971). A technique for determination of usual daily energy expenditure in the field. *American Journal of Clinical Nutrition,* **24,** 1148-1154.

Bradfield, R. (1979). Short term repeatability of oxygen consumption-heart rate relationships. *American Journal of Clinical Nutrition,* **32,** 1758-1759.

Bradfield, R.B., Chan, H., Bradfield, N.E., & Payne, P.R. (1971). Energy expenditures and heart rates of Cambridge boys at school. *American Journal of Clinical Nutrition,* **24,** 1461-1466.

Bradfield, R.B., Huntzicker, P.B., & Fruehan, G.J. (1969). Simultaneous comparison of respirometer and heart rate telemetry techniques as measures of human energy expenditure. *American Journal of Clinical Nutrition,* **22,** 696-700.

Bradfield, R.B., Paulos, J., & Grossman, L. (1971). Energy expenditure and heart rate of obese high school girls. *American Journal of Clinical Nutrition,* **24,** 1482-1488.

Ceesay, S.M., Prentice, A.M., Day, K.C., Murgatroyd, P.R., Goldberg, G.R., Scott, W., & Spurr, G.B. (1989). The use of heart rate monitoring in the estimation of energy expenditure: A validation study using indirect whole-body calorimetry. *British Journal of Nutrition,* **61,** 175-186.

Christensen, C.C., Frey, H.M.M., Foenstelien, E., Aadland, E., & Refsum, H.E. (1983). A critical evaluation of energy expenditure estimates based on individual O_2 consumption/heart rate curves and average daily heart rate. *American Journal of Clinical Nutrition,* **37,** 468-472.

Collins, K.J., & Spurr, G.B. (1990). Energy expenditure and habitual activity. In K.J. Collins (Ed.), *Handbook of methods for the measurement of work performance physical fitness and energy expenditure in tropical populations* (pp. 81-93). Paris: IUBS.

Collins, M.A., Cureton, K.J., Hill, D.W., & Ray, C.A. (1991). Relationship of heart rate to oxygen uptake during weight lifting exercise. *Medicine and Science in Sports and Exercise,* **23,** 636-640.

Corbin, C.B., & Pletcher, P. (1968). Diet and physical activity patterns of obese and non-obese elementary school children. *Research Quarterly,* **39,** 922-928.

Dauncy, M.J., & James, W.P.T. (1979). Assessment of the heart rate method for determining energy expenditure in man, using a whole-body calorimeter. *British Journal of Nutrition,* **42,** 1-13.

Davies, C.T.M., Tuxworth, W., & Young, J. (1970). Physiological effects of repeated exercise. *Clinical Science,* **39,** 247-258.

Durant, R.H., Baranowski, T., Davis H., Rhodes, T., Thompson, W.O., Greaves, K.A., & Puhl, J. (1993). Reliability and variability of indicators of heart rate monitoring in children. *Medicine and Science in Sports and Exercise,* **25,** 389-395.

Durant, R.H., Baranowski, T., Davis, H., Thompson, W.O., Puhl, J., Greaves, K.A., & Rhodes, T. (1992). Reliability and variability of heart rate monitoring in 3-, 4-, or 5-yr-old children. *Medicine and Science in Sports and Exercise,* **24,** 265-271.

Durin, N., & Namyslowski L. (1958). Individual variations in energy expenditure of standardized activities. *Journal of Physiology,* **143,** 573-578.

Durnin, J.V.G.A., & Edwards, R.G. (1955). Pulmonary ventilation as an index of energy expenditure. *Quarterly Journal of Experimental Physiology,* **40,** 370-377.

Edholm, D.G. (1966). The assessment of habitual activity in health and disease. *Proceedings of the Beitosolen Symposium* (pp. 187-197). Oslo: Universitetsforlaget.

Elfner, R., Buss, J., Heene, D.L., & Kraatz, J. (1988). Beat-by-beat validation of the Oxford Medilog 4500, a 24-hour ambulatory ECG system with real-time analysis. *Journal of Ambulatory Monitoring,* **1,** 17-31.

Emons, H.J.G., Groenenboom, D.C., Westerterp, K.R., & Saris, W.H.M. (1992). Comparison of heart rate monitoring combined with indirect calorimetry and the doubly labeled water ($^2H_2$ ^{18}O) method for the measurement of energy expenditure in children. *European Journal of Applied Physiology,* **65,** 99-103.

Erickson, S., Simonson, E., Taylor, H., Alexander, H., & Keys, A. (1945). The energy cost of horizontal and grade walking on the motor driven treadmill. *American Journal of Physiology,* **145,** 391-401.

Fordham, M., Goldsmith, R., Kaval, J., O'Brien, C., & Tan, G.L.E. (1978). Ambulatory monitoring in industry—heart rate or oxygen consumption as indices of strain. In F.D. Stott, E.B. Raftery, P. Sleight, & L. Golding (Eds.), *Proceedings of the 2nd International Symposium on Ambulatory Monitoring* (pp. 283-297). London: Academic Press.

Fouillot, J.P., Drozpowski, T., Tekaia, F., Regnard, J., Izou, M.A., Fourneron, T., Leblanc, A., & Rieu, M. (1982). Methodology of heart rate ambulatory monitoring recordings analysis, in relation to activity: Applications to sports training and workload studies. In F.D. Stott, E.B. Raftery, D.L. Clement, & S.L. Wright (Eds.), *Proceedings of the 4th International Symposium on Ambulatory Monitoring and the Second Gent Workshop on Blood Pressure Variability, 1981* (pp. 377-383). London: Academic Press.

Friedman, M., Rosenman, R.H., & Brown, A.E. (1963). The continuous heart rate in men exhibiting an overt behavior pattern associated with increased incidence of clinical coronary artery disease. *Circulation,* **28,** 861-866.

Gandra, Y.R., & Bradfield, R.B. (1971). Energy expenditure and oxygen handling efficiency of anemic school children. *American Journal of Clinical Nutrition,* **24,** 1451-1456.

Glagov, S., Rowley, D.A., Cramer, D.B., & Page, R.G. (1970). Heart rates during 24 hours of usual activity for 100 normal men. *Journal of Applied Physiology,* **29,** 799-805.

Goldberger, E. (1961). Long period continuous electrocardiography of active persons. *American Journal of Cardiology,* **8,** 603-604.

Goldsmith, R., & Hale, T. (1971). Relationship between habitual physical activity and physical fitness. *American Journal of Clinical Nutrition,* **24,** 1489-1493.

Gretebeck, R.J., & Montoye, H.J. (1992). Variability of some objective measures of physical activity. *Medicine and Science in Sports and Exercise,* **24,** 1167-1172.

Gretebeck, R.J., Montoye, H.J., Ballor, D., & Montoye, A.P. (1991). Comment on heart rate recording in field studies. *Journal of Sports Medicine and Physical Fitness,* **31,** 629-631.

Griffiths, M., & Payne, P.R. (1976). Energy expenditure in small children of obese and non-obese parents. *Nature,* **260,** 698-700.

Hanish, H.M., Neustein, R.A., Van Cott, C.C., & Sanders, R.T. (1971). Technical aspects of monitoring the heart rate of active persons. *American Journal of Clinical Nutrition,* **24,** 1155-1163.

Hansen, O.E., & Maggio, M. (1960). Static work and heart rate. *Internationales Zeitschrift für Angewandte Physiologie Einschliesslich Arbeitsphysiologie,* **18,** 242-247.

Haskell, W.L., Yee, M.C., Evans, A., & Irby, P.J. (1993). Simultaneous measurement of heart rate and body motion to quantitate physical activity. *Medicine and Science in Sports and Exercise,* **25,** 109-115.

Hebestreit, H., Zehr, P., McKinty, C., Riddell, M., & Bar-Or, O. (1993). Correcting children's heart rate for the influence of climatic heat stress: Equation for an improved estimation of energy expenditure [abstract]. *Pediatric Exercise Science,* **5,** 428.

Hellstrom, R., & Holmgren, A. (1966). On the repeatability of submaximal work tests and the influence of body position on heart rate during exercise at submaximal work loads. *Scandinavian Journal of Clinical Laboratory Investigation,* **18,** 479-485.

Henderson, Y., & Prince, A.L. (1914). The oxygen pulse and the systolic discharge. *American Journal of Physiology,* **35,** 106-115.

Heywood, P.F., & Latham, M.C. (1971). Use of the SAMI heart-rate integrator in malnourished children. *American Journal of Clinical Nutrition,* **24,** 1146-1450.

Holter, N.J. (1961). New method for heart studies. *Science,* **134,** 1214-1220.

Kalkwarf, H.J., Haas, J.D., Belko, A.Z., Roach, R.C., & Roe, D.A. (1989). Accuracy of heart rate monitoring and activity diaries for estimating energy expenditure. *American Journal of Clinical Nutrition,* **49,** 37-43.

Kappagoda, C., Linden, R., & Newell, J. (1979). Effects of the Canadian Air Force training programme on a submaximal exercise test. *Quarterly Journal of Experimental Physiology,* **64,** 185-204.

Karvonen, J., Chwalbinska-Moneta, J., & Saynajakangas, S. (1984). Comparison of heart rates measured by ECG and microcomputer. *Physician and Sports Medicine,* **12,** 65-69.

Kemper, H.C.G. (1992). How important is physical activity for the development of aerobic power in youth? *Science et Motricité,* **17,** 44-50.

Kemper, H.C.G., Van Aalst, R., Leegwater, A., Maas, S., & Knibbe, J.J. (1990). The physical and physiological workload of refuse collectors. *Ergonomics,* **33,** 1473-1486.

Léger, L., & Thivierge, M. (1988). Heart rate monitors: Validity, stability, and functionality. *Physician and Sports Medicine,* **16,** 143-151.

Livingston, M.B.E., Prentice, A.M., Coward, W.A., Ceesay, S.M., Strain, J.J., McKenna, P.G., Nevin, G.B., Barker, M.E., & Hickey, R.J. (1990). Simultaneous measurement of free-living energy expenditure by the double labeled water method and heart rate monitoring. *American Journal of Clinical Nutrition,* **52,** 59-65.

Lundgren, N.P.V. (1946). The physiological effects of time schedule work on lumber workers. *Acta Physiologica Scandinavica,* **13,** Suppl. 41.

Maas, S., Kok, M.L., Westra, H.G., & Kemper, H.C.G. (1989). The validity of the use of heart rate in estimating oxygen consumption in static and in combined static/dynamic exercise. *Ergonomics,* **32,** 141-148.

MacConnie, S.E., Gilliam, T.B., Geenan, D.L., & Pels, A.E., III (1982). Daily physical activity patterns of prepubertal children involved in a vigorous exercise program. *International Journal of Sports Medicine,* **3,** 202-207.

Malhotra, M.S., Sen Gupta, J., & Rai, R.M. (1963). Pulse count as a measure of energy expenditure. *Journal of Applied Physiology,* **18,** 994-996.

Mansourian, P., Masironi, R., Nicoud, J.N., & Steffen, P. (1975). Recording the cardiac interbeat interval distribution. *Journal of Applied Physiology,* **38,** 542-545.

Maxfield, M. (1971). The indirect measurement of energy expenditure in industrial situations. *American Journal of Clinical Nutrition,* **24,** 1126-1130.

Meijer, G.A.L., Westerterp, K.R., Koper, H. & ten Hoor, F. (1989). Assessment of energy expenditure by recording heart rate and body acceleration. *Medicine and Science in Sports and Exercise,* **21,** 343-347.

Miles, L.E., & Rule, R.B. (1981). Long-term monitoring of multiple physiological parameters using a programmable portable microcomputer. In F.D. Scott, E.B. Raftery, D.L. Clement, & S.L. Wright (Eds.)

Proceedings of the 4th International Symposium on Ambulatory Monitoring and the Second Gent Workshop on Blood Pressure Variability, 1981 (pp. 249-257). London: Academic Press.

Mimura, K., Hebestreit, H.. & Bar-Or, O. (1991). Activity and heart rate in preschool children of low and high motor ability: 24 hour profiles [abstract]. *Medicine and Science in Sports in Exercise,* **23,** S12.

Montoye, H.J. (1970). Circulatory-respiratory fitness. In H.J. Montoye (Ed.), *An introduction to measurement in physical education, Vol. 4* (pp. 41-87). Indianapolis, IN: Phi Epsilon Kappa Fraternity.

Montoye, H.J. (1975). *Physical activity and health: An epidemiologic study of a total community.* Englewood Cliffs, NJ: Prentice-Hall.

Montoye, H., Cunningham, D., Welch, H., & Epstein, F. (1970). Laboratory methods for assessing metabolic capacity in a large epidemiologic study. *American Journal of Epidemiology,* **91,** 38-47.

Morgan, W., & Pollock, M. (1978). Physical activity and cardiovascular health: Psychological aspects. In F. Landry & W. Orban (Eds.), *Physical activity and human well-being* (pp. 163-181). Miami: Symposia Specialists.

Mueller, J.K., Gossard, D., Adams, F.R.,Taylor, C.B., Haskell, W.L., Kraemer, H.C., Ahn, D.K., Burnett, K., & DeBusk, R.F. (1986). Assessment of prescribed increases in physical activity: Application of a new method for microprocessor analysis of hear rate. *American Journal of Cardiology,* **57,** 441-445.

Müller, E.A., & Reeh, J.J. (1950). Die fortlaufende Registrierung der Pulsfrequenz bei beruflicher Arbeit [Continuous measurement of heart rate during physical work]. *Arbeitsphysiologie,* **14,** 137-146.

Murlin, J.R., & Greer, J.R. (1914). The relation of heart action to respiratory metabolism. *American Journal of Physiology,* **33,** 253.

O'Hara, N.M., Baranowski, T., Simons-Morton, B.G., Wilson, B.S., & Parcel, G. (1989). Validity of the observation of children's physical activity. *Research Quarterly for Exercise and Sport,* **60,** 42-47.

Patrick, J.M., Bassey, E.J., Irving, J.M., Blecher, A., & Fentem, P.H. (1986). Objective measurements of customary physical activity in elderly men and women before and after retirement. *Quarterly Journal of Experimental Physiology,* **71,** 47-58.

Payne, P.R., Wheeler, E.F., & Salvosa, C.B. (1971). Prediction of daily energy expenditure from average pulse rate. *American Journal of Clinical Nutrition,* **24,** 1164-1170.

Purvis, J., & Morgan, W. (1978). Influence of repeated maximal testing on anxiety and work capacity in college women. *Research Quarterly for Exercise and Sport,* **49,** 512-519.

Rhyming, I. (1954). A modified Harvard step test for the evaluation of physical fitness. *Arbeitsphysiologie,* **16,** 235.

Richardson, J.F. (1971). Heart rate in middle-aged men. *American Journal of Clinical Nutrition,* **24,** 1476-1481.

Rowley, D.A., Glagov, S., & Stoner, P. (1959). Measurement of human heart rate during usual activity. *Science,* **130,** 976-977.

Rutenfranz, J., Seliger, V., Lange Andersen, K., Ilmarinen, J., Flöring, R., Rutenfranz, M., & Klimmer, F. (1977). Erfahrungen mit einen transportablen Gerät zur zeiten bis zu 24 Studen [Experiences with a portable apparatus to time up to 24 hours]. *European Journal of Applied Physiology,* **36,** 171-185.

Sallis, J.F., Buono, M.J., Roby, J.J., Carlson, D., & Nelson, J.A. (1990). The Caltrac accelerometer as a physical activity monitor for school-age children. *Medicine and Science in Sports and Exercise,* **22,** 698-703.

Salvosa, C.B., Payne, P.R., & Wheeler, E.F. (1971). Energy expenditure of elderly people living alone or in local authority homes. *American Journal of Clinical Nutrition,* **24,** 1467-1470.

Saris, W.H.M. (1986). Habitual physical activity in children: Methodology and findings in health and disease. *Medicine and Science in Sports and Exercise,* **18,** 253-263.

Saris, W.H.M., Baecke, J., & Binkhorst, R.A. (1982). Validity of the assessment of energy expenditure from heart rate. In *Aerobic power and daily physical activity in children* (pp. 100-117). Saris, W.H.M., PhD thesis, University of Nymegen, Meppel, The Netherlands: Krips-Repro.

Saris, W.H.M., Binkhorst, R.A., Cramwinckel, A.B., Van Der Veen-Hezemans, A.M., & Van Waesberghe, F. (1979). Evaluation of somatic effects of a health education program for school children. *Bibliotheca Nutritio et Dieta,* **27,** 77-84.

Saris, W.H.M., Boeyen, I., Elvers, H., De Boo, T., & Binkhorst, R.A. (1982). Determining the individual variation in energy metabolism in 8-year-old children by two prediction methods. *Nutrition Reports International,* **26,** 35-42.

Saris, W.H.M., Elvers, J.W.H., Van't Hof, M.A., & Binkhorst, R.A. (1986). Changes in physical activity of children 6 to 12 years. In J. Rutenfranz, R. Mocellin, & F. Klimt (Eds.), *Children and exercise XII* (pp. 121-130). Champaign, IL: Human Kinetics.

Saris, W.H.M., Noordeloos, A.M., Cramwinckel, A.B., Boeyen, I., Elvers, J.W.H., van Veen, M., König, K.G., & Binkhorst, R.A. (1982). Aerobic power, daily physical activity and some cardio-vascular disease risk indicators in children ages 6-10 years.

In *Aerobic power and daily physical activity in children* (pp. 153-176). Saris, W.H.M., PhD thesis, University of Nymegen, Meppel, The Netherlands: Krips-Repro.

Saris, W.H.M., Snel, P., Baecke, J., Van Waesberghe, F., & Binkhorst, R.A. (1977). A portable miniature solid-state heart rate recorder for monitoring daily physical activity. *Biotelemetry,* **4,** 131-140.

Saris, W.H.M., Snel, P., & Binkhorst, R.A. (1977). A portable heart rate distribution recorder for studying daily physical activity. *European Journal of Applied Physiology,* **37,** 17-27.

Schultz, S., Westerterp, K.R., & Brück, K. (1989). Comparison of energy expenditure by the doubly labeled water technique with energy intake, heart rate, and activity recording in man. *American Journal of Clinical Nutrition,* **49,** 1146-1154.

Shephard, R. (1969). Learning, habituation and training. *Internationale Zeitschrift für Angewandte Physiologie Einschliesslich Arbeitsphysiologie,* **28,** 38-40.

Sime, W., Whipple, I., Berkson, D., MacIntyre, W., & Stamler, J. (1972). Reproducibility of heart rate at rest and in response to submaximal treadmill and bicycle ergometer exercise test in middle-aged men. *Medicine and Science in Sports and Exercise,* **4,** 14-17.

Spurr, G.B., Prentice, A.M., Murgatroyd, P.R., Goldberg, G.R., Reina, J.C., & Christman, N.T. (1988). Energy expenditure from minute-by-minute heart rate recording: Comparison with indirect calorimetry. *American Journal of Clinical Nutrition,* **48,** 552-559.

Taylor, C.B., Kraemer, H., Bragg, D., Miles, L., Rule, B., Savin, M., & DeBusk, R. (1980). A new system for long-term recording and processing of heart rate and physical activity in outpatients [abstract]. *Circulation,* (Suppl. III), **62,** 288.

Taylor, C.B., Kraemer, H., Bragg, D.A., Miles, L.E., Rule, B., Savin, W.M., & DeBusk, R.F. (1982). A new system for long-term recording and processing of heart rate and physical activity in outpatients. *Computers and Biomedical Research,* **15,** 7-17.

Taylor, H.L. (1967). Occupational factors in the study of coronary heart disease and physical activity. *Canadian Medical Association Journal,* **96,** 825-831.

Taylor, H.L., & Parlin, R.W. (June, 1966). *The physical activity of railroad clerks and switchmen: Estimation of on-the-job caloric expenditure by time and task measurements and classification of recreational activity by a questionnaire.* Paper presented at "Three Days of Cardiology," Seattle, Washington.

Torún, B. (1984). Physiological measurements of physical activity among children under free-living conditions. In A.R. Liss (Ed.), *Energy intake and activity* (pp. 159-184). New York: Alan R. Liss.

Treiber, F.A., Musante, L., Hartdagan, S., Davis, H., Levy, M., & Strong, W.B. (1989). Validation of a heart rate monitor with children in laboratory and field settings. *Medicine and Science in Sports and Exercise,* **21,** 338-342.

Van Dale, D., Schoffelen, P.F.M., ten Hoor, F., & Saris, W.H.M. (1989). Effects of addition of exercise to energy restriction on 24-hour energy expenditure, sleeping metabolic rate and daily physical activity. *European Journal of Clinical Nutrition,* **43,** 441-451.

Verschuur, R., & Kemper, H.C.G. (1985). Habitual physical activity. In M. Hebbelinck and R. Shephard (Eds.), *Medicine and Sport Science: Vol. 20. Growth, health and fitness of teenagers: Longitudinal research in international perspective* (pp. 56-65). Basel: Kargen.

Viteri, F.E., Tourin, B., Galicia, J., & Herrera, E. (1971). Determining energy costs of agricultural activities by respirometer and energy balance techniques. *American Journal of Clinical Nutrition,* **24,** 1418-1430.

Vokac, Z., Bell, H., Bautz-Holter, E., & Rodahl, K. (1975). Oxygen uptake/heart rate relationship in leg and arm exercise, sitting and standing. *Journal of Applied Physiology,* **39,** 54-59.

Warnold, H., Carlgren, G., & Krotkiewski, M. (1978). Energy expenditure and body composition during weight reduction in hyperplastic obese women. *American Journal of Clinical Nutrition,* **31,** 750-763.

Warnold, T., & Lenner, A.R. (1977). Evaluation of the heart rate method to determine the daily energy expenditure in disease. A study in juvenile diabetics. *American Journal of Clinical Nutrition,* **30,** 304-315.

Washburn, R.A., & Montoye, H.J. (1985). Reliability of the heart rate response to submaximal upper and lower body exercise. *Research Quarterly for Exercise and Sport,* **56,** 166-169.

Washburn, R.A., & Montoye, H.J. (1986). Validity of heart rate as a measure of mean daily energy expenditure. *Exercise Physiology,* **2,** 161-172.

Chapter 9

CONCLUSIONS

Several developments in industrialized nations have emphasized the need for valid measures of habitual physical activity. There is a growing concern about chronic disease and medical problems related to our lifestyles, including the amount of exercise in which we engage. If a lack of exercise is a predisposing factor in such medical problems, technology can be assigned part of the blame in having reduced the exercise required in our occupational and leisure time activities. It is conceivable that, had we had more precise measures of habitual physical activity, investigations might have revealed an even closer association between regular exercise and health.

Accurate methods of assessing exercise patterns are needed for a variety of reasons. The relationship of habitual physical activity to health and disease should continue to be explored, mostly by epidemiological studies involving large populations. Validated methods are needed for descriptive epidemiology (describing the physical activity in various populations, different ages, etc.) and for analytical epidemiology (e.g., the relationship of exercise to health or disease). Caspersen (1989) suggests it may also be important to define the relationship between physical activity and other behaviors.

Documenting changes in activity habits over time requires developing physical activity profiles of representative samples of an entire country and subpopulations within it. Exercise is being prescribed in educational and medical settings, and methods are needed to monitor compliance with these recommendations. Some individuals wish to know the energy they expend, and practical methods would be useful for this purpose.

Physical activity is a complex phenomenon. It is characterized by intensity (strenuousness), duration, frequency (per day, per week, etc.), and the surrounding circumstances. One can expend energy either in short bursts of strenuous activity or in less intense exercise over a longer period. The effects of these two regimens on health or fitness may be quite different. If energy expenditure is to be estimated as a reflection of physical activity, body weight (size) must be taken into account; the energy expended by a large person may well be greater than that of a small person, regardless of the exercise involved.

Despite the increased sedentariness of many occupations, if populations of working people are to be studied, we must assess the activity involved on the job, as well as during leisure time and getting to or from work. It also appears clear that methods of assessing physical activity among young children, young and middle-aged adults, and the elderly will of necessity differ. The techniques of estimating physical activity will also vary with the purpose of the assessment.

In some instances, only an average is needed for a particular population. Or it may only be necessary to classify people into three or four groups on the basis of exercise habits. Alternatively, individual scores might be needed. The precision of the method will vary with these requirements.

Measurement of recent physical activity, such as during the past week, can be expected to be more accurately recalled than during the past year or over a lifetime. If a week's appraisal is to reflect habitual activity over the entire year, however, week-to-week or seasonal variation must be considered; the previous week's activity may be atypical. Physical activity varies with the days in the week in many populations, and this also must be considered.

In the laboratory or clinic, energy expenditure and physical activity can be measured precisely by measuring oxygen consumption, heat production, or energy consumption. Valid measures of physical activity and energy expenditure in the field are needed. The questionable validity of present methods may be disappointing. Progress along these lines has been hampered by the lack of a good criterion against which practical methods may be compared. The doubly labeled water method has turned out to be the prime candidate for such a criterion. However, this technique measures only one aspect of physical activity, average energy expenditure. Also, there is still controversy concerning details of the procedure, and it is an expensive method. But the near future will see a resolution of the remaining problems in the doubly labeled water method, and the cost of the procedure may decrease.

We are not deluded into believing the present field methods will provide an accurate and highly precise measure of habitual physical activity or energy expenditure. It has been estimated that even with meticulous application of existing techniques, measurements of daily energy expenditure cannot be expected to be more accurate than 10% to 15% (Belding, 1960). This estimate was made more than 30 years ago, and although methods have been improving in recent years, the assessment procedures are still not very precise. On the other hand, in most instances this degree of accuracy is adequate.

Final Evaluation Summary

Table 9.1 summarizes the characteristics of field methods of assessing physical activity. We hope you will find this useful in selecting a method for your particular needs; however, bear in mind that some of the techniques (movement recorders, for example) are in their infancy, and improvements should be expected.

Suggestions for Further Research

In conclusion, we would like to suggest some needed avenues of research.

1. There is a need to improve the validity and reproducibility of questionnaires and interviews for particular populations.

2. We also need to know how recall can be improved among various populations (children, elderly, various cultural and ethnic groups).

3. If data from collection devices (questionnaires/interviews, observations, diaries, etc.) are to be converted to estimates of energy expenditure, more research is needed on the energy cost of particular activities and for specific populations and age groups. Most energy cost data are based on measurements of young, male adults. Data are particularly needed for female adults, children, and the elderly.

4. Better ways of using questionnaires and diaries to assess the strenuousness of activities are needed.

5. Also needed are more accurate and less expensive instruments, such as portable accelerometers, worn by the subject for estimating activity and energy expenditure.

6. Research is needed on the validity of combining methods, such as heart rate and movement recording, or movement recording and questionnaire/interview.

7. Exercise often has an anaerobic component, particularly for children during leisure activity. Research is needed to assess the importance of this with respect to methods of estimating energy expenditure and physical activity.

REFERENCES

Belding, H.S. (1960). Subcommittee on methods for acquiring information on physical activity and body form. *American Journal of Public Health and the Nation's Health Supplement,* **50,** 52-66.

Caspersen, C.J. (1989). Physical activity epidemiology: Concepts, methods, and applications to exercise science. *Exercise and Sport Sciences Reviews,* **17,** 423-473.

Table 9.1 Summary Evaluation of Field Methods of Assessing Habitual Physical Activity and/or Energy Expenditure

	Job classification	Observation	Diary	Questionnaire/ interview	Pedometer	Electronic motion sensors	Accelerometers	Heart rate	Doubly labeled water
Age group (yr) *appropriate*									
Children (<13 yr)	No	Yes	No	No	Yes	Yes	Yes	Yes	Yes
Adolescents (13-14 yr)	No	Yes	Yes	Yes	Yes	Yes	Yes	Yes	Yes
Adults (20-64 yr)	Yes	Unknown	Yes	Yes	Yes	Yes	Yes	Yes	Yes
Elderly (65+ yr)	No	Unknown	Yes	Yes	Yes	Yes	Yes	Yes	Yes
Validity									
Average energy expenditure	Poor	Good	Poor to good	Fair	Fair	Fair	Fair to good	Fair	High
Intensity of expenditure	Poor	Fair	Poor to good	Fair	Poor	Fair	Fair	Good	Poor
Reproducibility[a]	Unknown	Good	Unknown	Fair to good	Fair	Fair	Fair	Fair	Unknown
Instrument reliability[a]	Unknown	Good	Unknown	Unknown	Poor	Good	Good	Good	—
Size of population (Small: <50; Large: 50+)	Small/large	Small	Small/large	Small/large	Small/large	Small	Small/large	Small	Small
Cost	Low	High	Mod.	Low/Mod.	Mod.	Mod./High	Mod./High	High	High
Specific activities	No	Yes	Yes	Yes	No	No	No	No	No
Subject effort	None	None	Great	Mod.	Little	Little	Little	Little	None
Affects behavior	No	Yes	Yes	No	Possibly	Possibly	Possibly	Possibly	No
Subject acceptability	Yes	Yes	Yes/No	Yes	Yes	Yes	Yes	Yes/No	Yes
Summary index[b]	G	EE	EE	EE	Steps/G	Counts/G	EE, counts	EE	EE

[a]Reproducibility refers to repeated observations at another time; instrument reliability refers to consistence of response at the same observation.
[b]G: In most instances, subjects can only be grouped in broad categories on the basis of physical activity; EE: energy expenditure (mJ, kcal) either directly or indirectly, through use of energy cost tables of activities or grouping of subjects into broad categories.

Abbreviations, Symbols, and Conversion Table

Abbreviations and Symbols

A: absolute temperature scale (also known as Kelvin scale)

a–v̄O$_2$diff: difference in oxygen content between arterio and mixed venous blood

acc: Accelerometer

act: activity

ATPD: ambient temperature, pressure, dry

ATPS: ambient temperature, pressure, saturated

av: average, arithmetic mean

BMR: basal metabolic rate

BP: blood pressure

BTPS: body temperature, pressure, saturated

BTU: British thermal unit

°C: temperature in degrees Celsius or centigrade

cal: calorie

calib: calibration

Caltrac: trade name for one portable accelerometer

cc: cubic centimeter

CHD: coronary heart disease

cm: centimeter

CO$_2$: carbon dioxide

ct: count

da: day

diff: difference

DLW: doubly labeled water

E: expired gas (for example, V_E = volume of expired gas)

ECG: electrocardiogram

EE: energy expenditure

EMG: electromyogram

ergom: ergometer

est: estimate or estimated

F: fractional concentration (for example, F_IO_2 = fraction of oxygen in inspired gas)

°F: temperature in degrees Fahrenheit

ft: foot

g: gram

gal: gallon

^{2}H: isotope of hydrogen, deuterium

HDL: high density lipoprotein

Hg: mercury

HIP: Health Insurance Plan of New York questionnaire

hr: hour

HR: heart rate

I: inspired gas (for example, V_I = inspired gas volume)

IMP: integrating motor pneumotachograph

in: inch

ind: individual

insig: statistically insignificant ($p > 0.05$)

int: interview

inter: interval

J: joule

k: kilo

K: Kelvin temperature scale (also known as absolute scale)

kcal: kilocalorie, or large calorie

kg: kilogram

kJ: kilojoule

km: kilometer

L: liter

lb: pound

LSI: Large-Scale Integrated Motor Activity Monitor

LT: leisure time

m: meter

MET: ratio of exercise metabolic rate to resting metabolic rate; 1 MET = resting metabolic rate, 5 METs = five times the resting metabolic rate.

mg: milligram

min: minute

Minn: Minnesota

ml: milliliter

mm: millimeter

mph: miles per hour

n: number of subjects in a subset of the total (*N*)

N: total number of subjects

NS: not statistically significant ($p > 0.05$)

O_2: oxygen

$^{18}O_2$: isotope of oxygen

occ: occupational

overest: overestimated

oz: ounce

p: probability

P: pressure

PA: physical activity

PE: physical education

pt: pint

PWC 170: physical working capacity at a heart rate of 170 beats per minute

Q: questionnaire

QRS complex: ventricular depolarization

qt: quart

r: Pearson product moment correlation coefficient

reg: regression or regression equation

RER: respiratory exchange ratio

rest: resting

rho: Spearman rank order correlation coefficient

RMR: resting metabolic rate

RQ: respiratory quotient

s: second

SD: standard deviation

SEE: standard error of estimate (also SE_{est})

SEM: standard error of the mean

SI: International System of Units

sig: statistically significant ($p < 0.05$)

STPD: standard temperature, pressure, dry

$S_{y \cdot x}$: standard error of variable on y axis, standard error of estimate

T: temperature

Tecum: Tecumseh, Michigan (refers to questionnaire/interview developed for an investigation in that town)

test-retest coefficient: Pearson product moment correlation coefficient comparing one administration of a test with a second administration (also known as reliability coefficient)

underest: underestimated

V: gas volume

$\dot{V}$: gas volume per unit time, usually liters $\cdot$ min^{-1}

$\dot{V}CO_2$: rate of carbon dioxide flow; usually CO_2 exhaled or produced per minute

VO_2: volume of oxygen consumed, oxygen utilization

$\dot{V}O_2$: rate of oxygen flow; usually O_2 intake per minute

$\dot{V}O_2max$: maximum oxygen consumption, maximum oxygen utilization, maximum metabolic rate

W: weight

W_{170}: work accomplished at a heart rate of 170 beats per minute

YPAS: Yale Physical Activity Survey

Conversion Table

Heat and energy

1 calorie (cal) = 4.186 kilojoules (kJ) = $3.968 \cdot 10^{-3}$ British thermal units (BTU)
1 joule (J) = 0.239 calories = $0.948 \cdot 10^{-3}$ BTU
1 kilocalorie (kcal) = 1000 cal = 4186 J = 4.186 kJ = 3.968 BTU
1 kJ = 1000 J = 239 cal = 0.239 kcal

Distance

1 centimeter (cm) = 0.3937 inches (in.)
1 foot (ft) = 30.48 cm
1 in. = 2,540 cm
1 kilometer (km) = 0.6214 miles
1 meter (m) = 39.37 in. = 3.28 ft
1 mile = 1.609 km
1 millimeter (mm) = 0.03937 in.

Weight

1 gram (g) = .0353 ounces (oz) = 0.00294 pounds (lb)
1 kilogram (kg) = 35.3 oz = 2.20 lb
1 oz = 28.33 g = 0.0283 kg
1 lb = 453.3 g = 0.453 kg

Volume

1 gallon (gal) = 3.78 liters (L)
1 L = 1.057 quarts (qt) = 2.114 pints (pt) = 0.264 gal
1 pt = 0.473 L
1 qt = 0.946 L

Procedures for Measuring Oxygen Consumption ($\dot{V}O_2$): Open Circuit Method

Detailed step-by-step procedures and precautions for measuring oxygen consumption at rest or during exercise are available in many exercise physiology texts and laboratory manuals. The brief description here is meant simply to give an appreciation of what is entailed in the method and an understanding of its theoretical concepts. The procedure described here employs a Douglas bag, but exhaled air may be collected in a calibrated Tissot gasometer, a meteorological balloon, or other suitable apparatus.

The subject is fitted with a mask or a mouthpiece and nose clip and breathes through a one-way valve. The intake side of the valve is open to ambient air. The exhalation side is connected via rubber tubing about 2 in. in diameter to a three-way valve. By means of this valve, exhaled air from the subject may be directed to the room (or open air) or to the Douglas bag (when exhaled air is being collected). Ambient temperature (°C), barometric pressure (mm Hg), and relative humidity are recorded.

The exhaled air is collected in the Douglas bag for a precise period (measured with a stopwatch). One or more samples are extracted from the bag over mercury or in oiled syringes for later analysis by a Haldane or Micro-Scholander apparatus or by means of an electronic analyzer. The volume of the remaining gas in the Douglas bag is measured with a calibrated meter or Tissot gasometer and is added to the volume of the samples. The temperature of this gas is recorded to the closest 0.1 °C. The recorded data are then substituted in appropriate formulas to calculate oxygen consumption and carbon dioxide production, as detailed in the next section.

Oxygen Consumption Calculation

Oxygen consumption is calcualted by subtracting the amount of oxygen exhaled from the oxygen inhaled. Thus,

$$VO_2 = V_IO_2 - V_EO_2. \qquad (1)$$

When VO_2 is divided by the time period of observation, $\dot{V}O_2$ (usually per minute) is the result. V_IO_2 is calculated by multiplying the volume of air inhaled during the time period by the fraction (F_IO_2) of oxygen in the inhaled air. This is usually assumed to be 0.2093 (20.93%) if the subject is breathing outside or room air. The volume of expired oxygen (V_EO_2) is calculated by multiplying the volume of exhaled gas (V_E) by the fraction of oxygen (F_EO_2) in the exhaled gas. Thus,

$$VO_2 = (F_IO_2 \cdot V_I) - (F_EO_2 \cdot V_E) \qquad (2)$$

As mentioned, F_EO_2 is measured by a chemical analyzer (usually Haldane or Micro-Scholander) or an electronic analyzer. It is generally not necessary to measure both V_I and V_E because one can be calculated from the other. Usually, V_E is measured as already described. Because N_2 is not usually absorbed, N_2 inhaled must equal N_2 exhaled, so V_I can be calculated as follows:

$$V_I = \frac{(V_E \cdot F_EN_2)}{F_IN_2} \qquad (3)$$

If the subject is breathing room air, the fraction of nitrogen inhaled (F_IN_2) is usually assumed to be 0.7904 (79.04%). F_EN_2 is calculated by subtracting the fraction of the exhaled air that is oxygen (F_EO_2), measured by chemical or electronic analysis of the sample of exhaled air, and also by subtracting the fraction of exhaled air that is carbon dioxide (F_ECO_2), also measured by analysis of the sample of exhaled air. Thus, the equation for calculating F_EN_2 becomes

$$F_EN_2 = 1 - (F_EO_2 + F_ECO_2) \qquad (4)$$

Because the volume of exhaled air is measured, F_EN_2 is calculated by Formula 4, and F_IN_2 is assumed to be 0.7904, V_I can be calculated by Formula 3. Oxygen consumption can then be calculated by Formula 2, because all the values in the right-hand side of the equation are known.

Carbon Dioxide Production Calculation

The volume of carbon dioxide produced (VCO_2) can be calculated in a method similar to the calculation of oxygen.

$$VCO_2 = V_ECO_2 - V_ICO_2 \quad (5)$$

The volume of exhaled CO_2 (V_ECO_2) is calculated by multiplying the measured V_E by the fraction of CO_2 in exhaled air (measured by analysis of a sample of exhaled air). The volume of inhaled CO_2 (V_ICO_2) is calculated by multiplying the volume of inhaled air (Formula 3) by the fraction of CO_2 in the inhaled air (assumed to be 0.0003 [0.03%]).

Correction to Standard Conditions (STPD)

Gas volumes in the laboratory or the field are dependent on the ambient temperature and pressure and exhaled air is usually saturated (ATPS). Therefore, in order to standardize conditions, it is necessary to invoke Charles' Law and Boyles' Law and correct for water vapor. This is done as follows:

$$V_{STPD} = V_{ATPS}\left[\frac{273°}{(273° + T_A)}\right]\left[\frac{(P_A - P_{H_2O})}{P_S}\right] \quad (6)$$

where

T_A	=	ambient temperature in centigrade,
P_A	=	ambient barometric pressure in mm Hg,
P_{H_2O}	=	partial pressure of water vapor at ambient temperature (see Table B.1),
P_S	=	standard barometric pressure (760 mm Hg), and
$(273° + T_A)$		is the absolute temperature (273 °K) plus ambient temperature (°C).

Correction to Body Conditions

Ventilatory volumes are generally expressed as body temperature, pressure, saturated (BTPS). To correct volumes to BTPS, the following formula is employed:

Table B.1 Saturated Water Vapor Pressure at Various Temperatures

Temperature (°C)	Saturated water vapor pressure (mm Hg)
10	9.2
11	9.8
12	10.5
13	11.2
14	12.0
15	12.8
16	13.6
17	14.5
18	15.5
19	16.5
20	17.5
21	18.7
22	19.8
23	21.1
24	22.4
25	23.8
26	25.2
27	26.7
28	28.3
29	30.0
30	31.8
31	33.7
32	35.7
33	37.7
34	39.9
35	42.2
36	44.6
37	47.0

$$V_{BTPS} = V_{ATPS}\left[\frac{310°}{(273° + T_A)}\right]\left[\frac{(P_A - P_{H_2O})}{(P_S - 47)}\right]$$

where T_A, P_A, P_{H_2O}, and P_S are defined as for Formula 6. The figure 310° is derived from 273 °K + normal core temperature (37 °C). The figure 47 is water vapor pressure (47 mm Hg) at normal body temperature of 37 °C.

Appendix C

Classification by Energy Cost of Human Physical Activities

Many of the values of this list and those added at the end came from the following sources: Bannister and Brown (1968); the 7-Day Recall Physical Activity Questionnaire (Blair et al., 1985); Durnin and Passmore (1967); Howley and Glover (1974); the American Health Foundation's Physical Activity List (Leon, 1981); McArdle, Katch, and Katch (1988); Passmore and Durnin (1955); Tecumseh Questionnaire (Reiff et al., 1967a, 1967b). Some values have been added from the following sources: Collins, Cureton, Hill, and Ray (1991); Geissler et al. (1981); Getchell (1968); Goff, Frassetto, and Specht (1956); Mandli, Hoffman, Jones, Bota, and Clifford (1989); Nelson, Pells, Geenen, and White (1988); Seliger (1968); Stray-Gundersen and Galanes (1991); Veicsteinas, Ferretti, Margonato, Rosa, and Tagliabue (1984); VonHofen, Auble, and Schwartz (1989); Watts, Martin, Schmeling, Silta, and Watts (1990); Wigaeus and Kilbom (1980).

Much of the data in this appendix is derived from actual measurement by indirect calorimetry. However, where data are not available, the figures are based on educated guesses. For some activities, the values are not the values obtained exclusively during execution of the activities. For example, folk dancing requires a higher value than that shown. However, in an hour of folk dancing, considerable time is spent standing, receiving directions, and so on, so the value shown represents the estimated average value. On the other hand, walking usually is done continuously, so its values represent the actual energy cost of doing the activity.

Adults (usually young adults) served as subjects in determining almost all of the metabolic costs of activities that have been reported in the literature. Little data is based on children and the elderly. The energy expended by children in kilocalories per kilogram of body weight in performing even common activities such as walking is significantly higher than when the same activities are done by adults (Montoye, 1982). This is probably because of children's greater ratio of surface area to body weight and poorer coordination than adults. Even if the resting energy expenditure is also higher in children, the MET values of activities in the table are probably a little low for children. Data from Torún (1983) have shown the same results. This has also been shown to be true for infants (Torún, Chew, & Mendoza, 1983). Data on energy cost of activities are needed to create a table for children like the one in this appendix.

Data on the energy cost of elderly adults are also needed. Although walking at the same rate may elicit an energy expenditure not much different than in young adults, the elderly generally walk slower, play tennis at less intensity, skate less vigorously, and the like, so the estimate of habitual energy expenditure in the elderly requires other energy cost values.

The numerical value on the left is the MET rating (the energy cost of the activity divided by resting energy expenditure). This MET value is approximately equal to the energy cost of the activity, expressed as kilocalories per hour per kilogram of body weight. For more detail on how to apply the values in the list, see Ainsworth et al., 1993.

METs	Activity category	Specific activity
8.5	Bicycling	Bicycling, BMX or mountain
4.0	Bicycling	Bicycling, <10 mph, general, leisure, to work or for pleasure
6.0	Bicycling	Bicycling, 10-11.9 mph, leisure, slow, light effort
8.0	Bicycling	Bicycling, 12-13.9 mph, leisure, moderate effort
10.0	Bicycling	Bicycling, 14-15.9 mph, racing or leisure, fast, vigorous effort
12.0	Bicycling	Bicycling, 16-19 mph, racing/not drafting or >19 mph drafting, very fast, racing general
16.0	Bicycling	Bicycling, >20 mph, racing, not drafting
5.0	Bicycling	Unicycling
5.0	Conditioning exercise	Bicycling, stationary, general
3.0	Conditioning exercise	Bicycling, stationary, 50 W, very light effort
5.5	Conditioning exercise	Bicycling, stationary, 100 W, light effort
7.0	Conditioning exercise	Bicycling, stationary, 150 W, moderate effort
10.5	Conditioning exercise	Bicycling, stationary, 200 W, vigorous effort
12.5	Conditioning exercise	Bicycling, stationary, 250 W, very vigorous effort
8.0	Conditioning exercise	Calisthenics (e.g., pushups, pullups, situps), heavy, vigorous effort
4.5	Conditioning exercise	Calisthenics, home exercise, light or moderate effort, general (example: back exercises), going up & down from floor
8.0	Conditioning exercise	Circuit training, general
6.0	Conditioning exercise	Weight lifting (free weight, nautilus or univeral-type), power lifting or body building, vigorous effort
5.5	Conditioning exercise	Health club exercise, general
6.0	Conditioning exercise	Stair-treadmill ergometer, general
9.5	Conditioning exercise	Rowing, stationary ergometer, general
3.5	Conditioning exercise	Rowing, stationary, 50 W, light effort
7.0	Conditioning exercise	Rowing, stationary, 100 W, moderate effort
8.5	Conditioning exercise	Rowing, stationary, 150 W, vigorous effort
12.0	Conditioning exercise	Rowing, stationary, 200 W, very vigorous effort
9.5	Conditioning exercise	Ski machine, general
6.0	Conditioning exercise	Slimnastics
4.0	Conditioning exercise	Stretching, hatha yoga
6.0	Conditioning exercise	Teaching aerobic exercise class
4.0	Conditioning exercise	Water aerobics, water calisthenics
3.0	Conditioning exercise	Weight lifting (free, nautilus or universal-type), light or moderate effort, light workout, general
1.0	Conditioning exercise	Whirlpool, sitting
6.0	Dancing	Aerobic, ballet or modern, twist
6.0	Dancing	Aerobic, general
5.0	Dancing	Aerobic, low impact
7.0	Dancing	Aerobic, high impact
4.5	Dancing	General
5.5	Dancing	Ballroom, fast (disco, folk, square)
3.0	Dancing	Ballroom, slow (e.g., waltz, foxtrot, slow dancing)
5.0	Fishing and hunting	Fishing, general
4.0	Fishing and hunting	Digging worms, with shovel
5.0	Fishing and hunting	Fishing from river bank and walking
2.5	Fishing and hunting	Fishing from boat, sitting
3.5	Fishing and hunting	Fishing from river bank, standing
6.0	Fishing and hunting	Fishing in stream, in waders
2.0	Fishing and hunting	Fishing, ice, sitting
2.5	Fishing and hunting	Hunting, bow and arrow or crossbow
6.0	Fishing and hunting	Hunting, deer, elk, large game
2.5	Fishing and hunting	Hunting, duck, wading
5.0	Fishing and hunting	Hunting, general
6.0	Fishing and hunting	Hunting, pheasants or grouse
5.0	Fishing and hunting	Hunting, rabbit, squirrel, prairie chick, raccoon, small game

2.5	Fishing and hunting	Pistol shooting or trap shooting, standing
2.5	Home activities	Carpet sweeping, sweeping floors
4.5	Home activities	Cleaning, heavy or major (e.g., wash car, wash windows, mop, clean garage), vigorous effort
3.5	Home activities	Cleaning, house or cabin, general
2.5	Home activities	Cleaning, light (dusting, straightening up, vacuuming, changing linen, carrying out trash), moderate effort
2.3	Home activities	Wash dishes-standing or in general (not broken into stand/walk components)
2.3	Home activities	Wash dishes; clearing dishes from table-walking
2.5	Home activities	Cooking or food preparation-standing or sitting or in general (not broken into stand/ walk components)
2.5	Home activities	Serving food, setting table-implied walking or standing
2.5	Home activities	Cooking or food preparation-walking
2.5	Home activities	Putting away groceries (e.g., carrying groceries, shopping without a grocery cart)
8.0	Home activities	Carrying groceries upstairs
3.5	Home activities	Food shopping, with grocery cart
2.0	Home activities	Standing-shopping (non-grocery shopping)
2.3	Home activities	Walking-shopping (non-grocery shopping)
2.3	Home activities	Ironing
1.5	Home activities	Sitting, knitting, sewing, light wrapping (presents)
2.0	Home activities	Implied standing-laundry, fold or hang clothes, put clothes in washer or dryer, packing suitcase
2.3	Home activities	Implied walking-putting away clothes, gathering clothes to pack, putting away laundry
2.0	Home activities	Making bed
5.0	Home activities	Maple syruping/sugar bushing (including carrying buckets, carrying wood)
6.0	Home activities	Moving furniture, household
5.5	Home activities	Scrubbing floors, on hands and knees
4.0	Home activities	Sweeping garage, sidewalk or outside of house
7.0	Home activities	Moving household items, carrying boxes
3.5	Home activities	Standing-packing/unpacking boxes, occasional lifting of household items, light-moderate effort
3.0	Home activities	Implied walking-putting away household items-moderate effort
9.0	Home activities	Move household items upstairs, carrying boxes or furniture
2.5	Home activities	Standing-light (pump gas, change light bulb, etc.)
3.0	Home activities	Walking-light, noncleaning (ready to leave, shut/lock doors, close windows, etc.)
2.5	Home activities	Sitting-playing with child(ren)-light
2.8	Home activities	Standing-playing with child(ren)-light
4.0	Home activities	Walk/run-playing with child(ren)-moderate
5.0	Home activities	Walk/run-playing with child(ren)-vigorous
3.0	Home activities	Child care: sitting/kneeling-dressing, bathing, grooming, feeding, occasional lifting of child-light effort
3.5	Home activities	Child care: standing-dressing, bathing, grooming, feeding, occasional lifting of child-light effort
3.0	Home repair	Airplane repair
4.5	Home repair	Automobile body work
3.0	Home repair	Automobile repair
3.0	Home repair	Carpentry, general, workshop
6.0	Home repair	Carpentry, outside house, installing rain gutters
4.5	Home repair	Carpentry, finishing or refinishing cabinets or furniture
7.5	Home repair	Carpentry, sawing hardwood
5.0	Home repair	Caulking, chinking log cabin
4.5	Home repair	Caulking, except log cabin
5.0	Home repair	Cleaning gutters
5.0	Home repair	Excavating garage
5.0	Home repair	Hanging storm windows
4.5	Home repair	Laying or removing carpet
4.5	Home repair	Laying tile or linoleum
5.0	Home repair	Painting, outside house
4.5	Home repair	Painting, papering, plastering, scraping, inside house, hanging sheet rock, remodeling

METs	Activity category	Specific activity
3.0	Home repair	Put on and removal of tarp-sailboat
6.0	Home repair	Roofing
4.5	Home repair	Sanding floors with a power sander
4.5	Home repair	Scrape and paint sailboat or power boat
5.0	Home repair	Spreading dirt with a shovel
4.5	Home repair	Wash and wax hull of sailboat, car, powerboat, airplane
4.5	Home repair	Washing fence
3.0	Home repair	Wiring, plumbing
0.9	Inactivity, quiet	Lying quietly, reclining (watch television), lying quietly in bed-awake
1.0	Inactivity, quiet	Sitting quietly (riding in a car, listening to a lecture or music, watch television or a movie)
0.9	Inactivity, quiet	Sleeping
1.2	Inactivity, quiet	Standing quietly (standing in a line)
1.0	Inactivity, light	Recline-writing
1.0	Inactivity, light	Recline-talking or talking on phone
1.0	Inactivity, light	Recline-reading
5.0	Lawn and garden	Carrying, loading or stacking wood, loading/unloading or carrying lumber
6.0	Lawn and garden	Chopping wood, splitting logs
5.0	Lawn and garden	Clearing land, hauling branches
5.0	Lawn and garden	Digging sandbox
5.0	Lawn and garden	Digging, spading, filling garden
6.0	Lawn and garden	Gardening with heavypower tools, tilling a garden (see occupation, shoveling)
5.0	Lawn and garden	Laying crushed rock
5.0	Lawn and garden	Laying sod
5.5	Lawn and garden	Mowing lawn, general
2.5	Lawn and garden	Mowing lawn, riding mower
6.0	Lawn and garden	Mowing lawn, walk, hand mower
4.5	Lawn and garden	Mowing lawn, walk, power mower
4.5	Lawn and garden	Operating snow blower, walking
4.0	Lawn and garden	Planting seedlings, shrubs
4.5	Lawn and garden	Planting trees
4.0	Lawn and garden	Raking lawn
4.0	Lawn and garden	Raking roof with snow rake
3.0	Lawn and garden	Riding snow blower
4.0	Lawn and garden	Sacking grass, leaves
6.0	Lawn and garden	Shoveling, snow, by hand
4.5	Lawn and garden	Trimming shrubs or trees, manual cutter
3.5	Lawn and garden	Trimming shrubs or trees, power cutter
2.5	Lawn and garden	Walking, applying fertilizer or seeding a lawn
1.5	Lawn and garden	Watering lawn or garden, standing or walking
4.5	Lawn and garden	Weeding, cultivating garden
5.0	Lawn and garden	Gardening, general
3.0	Lawn and garden	Implied walking/standing-picking up yard, light
1.5	Miscellaneous	Sitting, card playing, playing board games
2.0	Miscellaneous	Standing-drawing (writing), casino gambling
1.3	Miscellaneous	Sitting-reading, book, newspaper, etc.
1.8	Miscellaneous	Sitting-writing, desk work
1.8	Miscellaneous	Standing-talking or talking on the phone
1.5	Miscellaneous	Sitting-talking or talking on the phone
1.8	Miscellaneous	Sitting-studying, general, including reading and/or writing
1.8	Miscellaneous	Sitting-in class, general, including note-taking or class discussion
1.8	Miscellaneous	Standing-reading
1.8	Music playing	Accordion
2.0	Music playing	Cello
2.5	Music playing	Conducting
4.0	Music playing	Drums

2.0	Music playing	Flute (sitting)
2.0	Music playing	Horn
2.5	Music playing	Piano or organ
3.5	Music playing	Trombone
2.5	Music playing	Trumpet
2.5	Music playing	Violin
2.0	Music playing	Woodwind
2.0	Music playing	Guitar, classial, folk (sitting)
3.0	Music playing	Guitar, rock and roll band (standing)
4.0	Music playing	Marching band, playing an instrument, baton twirling (walking)
3.5	Music playing	Marching band, drum major (walking)
4.0	Occupation	Bakery, general
2.3	Occupation	Bookbinding
6.0	Occupation	Building road (including hauling debris, driving heavy machinery)
2.0	Occupation	Building road, directing traffic (standing)
3.5	Occupation	Carpentry, general
8.0	Occupation	Carrying heavy loads, such as bricks
8.0	Occupation	Carrying moderate loads up stairs, moving boxes (16-40 pounds)
2.5	Occupation	Chambermaid
6.5	Occupation	Coal mining, drilling coal, rock
6.5	Occupation	Coal mining, erecting supports
6.0	Occupation	Coal mining, general
7.0	Occupation	Coal mining, shoveling coal
5.5	Occupation	Construction, outside, remodeling
3.5	Occupation	Electrical work, plumbing
8.0	Occupation	Famring, baling hay, cleaning barn, poultry work
3.5	Occupation	Farming, chasing cattle, nonstrenuous
2.5	Occupation	Farming, driving harvester
2.5	Occupation	Farming, driving tractor
4.0	Occupation	Farming, feeding small animals
4.5	Occupation	Farming, feeding cattle
8.0	Occupation	Farming, forking straw bales
3.0	Occupation	Farming, milking by hand
1.5	Occupation	Farming, milking by machine
5.5	Occupation	Farming, shoveling grain
12.0	Occupation	Fire fighter, general
11.0	Occupation	Fire fighter, climbing ladder with full gear
8.0	Occupation	Fire fighter, hauling hoses on ground
17.0	Occupation	Forestry, ax chopping, fast
5.0	Occupation	Forestry, ax chopping, slow
7.0	Occupation	Forestry, barking trees
11.0	Occupation	Forestry, carrying logs
8.0	Occupation	Forestry, felling trees
8.0	Occupation	Forestry, general
5.0	Occupation	Forestry, hoeing
6.0	Occupation	Forestry, planting by hand
7.0	Occupation	Forestry, sawing by hand
4.5	Occupation	Forestry, sawing, power
9.0	Occupation	Forestry, trimming trees
4.0	Occupation	Forestry, weeding
4.5	Occupation	Furriery
6.0	Occupation	Horse grooming
8.0	Occupation	Horse racing, galloping
6.5	Occupation	Horse racing, trotting
2.6	Occupation	Horse racing, walking
3.5	Occupation	Locksmith
2.5	Occupation	Maching tooling, machining, working sheet metal
3.0	Occupation	Machine tooling, operating lathe
5.0	Occupation	Machine tooling, operating punch press

METs	Activity category	Specific activity
4.0	Occupation	Machine tooling, tapping and drilling
3.0	Occupation	Maching tooling, welding
7.0	Occupation	Masonry, concrete
4.0	Occupation	Masseur, masseuse (standing)
7.0	Occupation	Moving, pushing heavy objects, 75 lbs or more (desks, moving van work)
2.5	Occupation	Operating heavy duty euqipment/automated, not driving
4.5	Occupation	Orange grove work
2.3	Occupation	Printing (standing)
2.5	Occupation	Police, directing traffic (standing)
2.0	Occupation	Police, driving a squad car (sitting)
1.3	Occupation	Police, riding in a squad car (sitting)
8.0	Occupation	Police, making an arrest (standing)
2.5	Occupation	Shoe repair, general
8.5	Occupation	Shoveling, digging ditches
9.0	Occupation	Shoveling, heavy (more than 16 lbs $\cdot$ min^{-1})
6.0	Occupation	Shoveling, light (less than 10 lbs $\cdot$ min^{-1})
7.0	Occupation	Shoveling, moderate (10-15 lbs $\cdot$ min^{-1})
1.5	Occupation	Sitting-light office work, in general (chemistry lab work, light use of handtools, watch repair or micro-assembly, light assembly/repair)
1.5	Occupation	Sitting-meetings, general, and/or with talking involved
2.5	Occupation	Sitting; moderate (heavy levers, riding mower/forklift, crane operation)
2.5	Occupation	Standing; light (bartending, store clerk, assembling, filing, xeroxing, put up Christmas tree)
3.0	Occupation	Standing; light/moderate (assemble/repair heavy parts, welding, stocking, auto repair, pack boxes for moving, etc.), patient care (as in nursing)
3.5	Occupation	Standing; moderate (assembling at fast rate, lifting 50 lbs, hitch/twisting ropes)
4.0	Occupation	Standing; moderate/heavy (lifting more than 50 lb, masonry, painting, paper hanging)
5.0	Occupation	Steel mill, fettling
5.5	Occupation	Steel mill, forging
8.0	Occupation	Steel mill, hand rolling
8.0	Occupation	Steel mill, merchant mill rolling
11.0	Occupation	Steel mill, removing slag
7.5	Occupation	Steel mill, tending furnace
5.5	Occupation	Steel mill, tipping molds
8.0	Occupation	Steel mill, working in general
2.5	Occupation	Tailoring, cutting
2.5	Occupation	Tailoring, general
2.0	Occupation	Tailoring, hand sewing
2.5	Occupation	Tailoring, machine sewing
4.0	Occupation	Tailoring, pressing
6.5	Occupation	Truck driving, loading and unloading truck (standing)
1.5	Occupation	Typing, electric, manual or computer
6.0	Occupation	Using heavy power tools such as pneumatic tools (jackhammers, drills, etc.)
8.0	Occupation	Using heavy tools (not power) such as shovel, pick, tunnel bar, spade
2.0	Occupation	Walking on job, less than 2.0 mph (in office or lab area), very slow
3.5	Occupation	Walking on job, 3.0 mph, in office, moderate speed, not carrying anything
4.0	Occupation	Walking on job, 3.5 mph, in office, brisk speed, not carrying anything
3.0	Occupation	Walking, 2.5 mph, slowly and carrying light objects less than 25 lbs
4.0	Occupation	Walking, 3.0 mph, moderately and carrying light objects less than 25 lbs
4.5	Occupation	Walking, 3.5 mph, briskly and carrying objects less than 25 lbs
5.0	Occupation	Walking or walk downstairs or standing, carrying objects about 25–49 lbs
6.5	Occupation	Walking or walk downstairs or standing, carrying objects about 50–74 lbs
7.5	Occupation	Walking or walk downstairs or standing, carrying objects about 75–99 lbs
8.5	Occupation	Walking or walk downstairs or standing, carrying objects about 100 lbs and over
3.0	Occupation	Working in scene shop, theater actor, backstage, employee
6.0	Running	Job/walk combination (jobbing component of less than 10 min)

7.0	Running	Jogging, general
8.0	Running	Running, 5 mph (12 min · mile^{-1})
9.0	Running	Running, 5 mph (12 min · mile^{-1})
8.0	Running	Running, 5.2 mph (11.5 min · mile^{-1})
10.0	Running	Running, 6 mph (10 min · mile^{-1})
11.0	Running	Running, 6.7 mph (9 min · mile^{-1})
11.5	Running	Running, 7 mph (8.5 min · mile^{-1})
12.5	Running	Running, 7.5 mph (8 min · mile^{-1})
13.5	Running	Running, 8 mph (7.5 min · mile^{-1})
14.0	Running	Running, 8.6 mph (7 min · mile^{-1})
15.0	Running	Running, 9 mph (6.5 min · mile^{-1})
16.0	Running	Running, 10 mph (6 min · mile^{-1})
18.0	Running	Running, 10.9 mph (5.5 min · mile^{-1})
9.0	Running	Running, cross-country
8.0	Running	Running, general
8.0	Running	Running, in place
15.0	Running	Running, stairs, up
10.0	Running	Running, on a track, team practice
8.0	Running	Running, training, pushing wheelchair, marathon wheeling
3.0	Running	Running, wheeling, general
2.5	Self-care	Standing-getting ready for bed, in general
1.0	Self-care	Sitting on toilet
2.0	Self-care	Bathing (sitting)
2.5	Self-care	Dressing, undressing (standing or sitting)
1.5	Self-care	Eating (sitting)
2.0	Self-care	Talking and eating or eating only (standing)
2.5	Self-care	Sitting or standing-grooming (washing, shaving, brushing teeth, urinating, washing hands, put on make-up)
4.0	Self-care	Showering, toweling off (standing)
1.5	Sexual activity	Active, vigorous effort
1.3	Sexual activity	General, moderate effort
1.0	Sexual activity	Passive, light effort, kissing, hugging
3.5	Sports	Archery (nonhunting)
7.0	Sports	Badminton, competitive
4.5	Sports	Badminton, social singles and doubles, general
8.0	Sports	Basketball, game
6.0	Sports	Basketball, nongame, general
7.0	Sports	Basketball, officiating
4.5	Sports	Basketball, shooting baskets
6.5	Sports	Basketball, wheelchair
2.5	Sports	Billiards
3.0	Sports	Bowling
12.0	Sports	Boxing, in ring, general
6.0	Sports	Boxing, punching bag
9.0	Sports	Boxing, sparring
7.0	Sports	Broomball
5.0	Sports	Children's games (hopscotch, 4-square, dodgeball, playground apparatus, t-ball, tetherball, marbles, jacks, arcard games)
4.0	Sports	Coaching: football, soccer, basketball, baseball, swimming, etc.
5.0	Sports	Cricket (batting, bowling)
2.5	Sports	Croquet
4.0	Sports	Curling
2.5	Sports	Darts, wall or lawn
6.0	Sports	Drag racing, pushing or driving a car
6.0	Sports	Fencing
9.0	Sports	Football, competitive
8.0	Sports	Football, touch, flag, general
2.5	Sports	Football or baseball, playing catch
3.0	Sports	Frisbee playing, general

METs	Activity category	Specific activity
3.5	Sports	Frisbee, ultimate
4.5	Sports	Golf, general
5.5	Sports	Golf, carrying clubs
3.0	Sports	Golf, miniature, driving range
5.0	Sports	Golf, pulling clubs
3.5	Sports	Golf, using power cart
4.0	Sports	Gymnastics, general
4.0	Sports	Hacky sack
12.0	Sports	Handball, general
8.0	Sports	Handall, team
3.5	Sports	Hang gliding
8.0	Sports	Hockey, field
8.0	Sports	Hockey, ice
4.0	Sports	Horseback riding, general
3.5	Sports	Horseback riding, saddling horse
6.5	Sports	Horseback riding, trotting
2.5	Sports	Horseback riding, walking
3.0	Sports	Horseshoe pitching, quoits
12.0	Sports	Jai alai
10.0	Sports	Judo, jujitsu, karate, kick boxing, tae kwan do
4.0	Sports	Juggling
7.0	Sports	Kickball
8.0	Sports	Lacrosse
4.0	Sports	Moto-cross
9.0	Sports	Orienteering
10.0	Sports	Paddleball, competitive
6.0	Sports	Paddleball, casual, general
8.0	Sports	Polo
10.0	Sports	Racketball, competitive
7.0	Sports	Racketball, casual, general
11.0	Sports	Rock climbing, ascending rock
8.0	Sports	Rock climbing, rapelling
12.0	Sports	Rope jumping, fast
10.0	Sports	Rope jumping, moderate, general
8.0	Sports	Rope jumping, slow
10.0	Sports	Rugby
3.0	Sports	Shuffleboard, lawn bowling
5.0	Sports	Skateboarding
7.0	Sports	Skating, roller
3.5	Sports	Sky diving
10.0	Sports	Soccer, competitive
7.0	Sports	Soccer, casual, general
5.0	Sports	Softball or baseball, fast or slow pitch, general
4.0	Sports	Softball, officiating
6.0	Sports	Softball, pitching
12.0	Sports	Squash
4.0	Sports	Table tennis, ping pong
4.0	Sports	Tai chi
7.0	Sports	Tennis, general
6.0	Sports	Tennis, doubles
8.0	Sports	Tennis, singles
3.5	Sports	Trampoline
4.0	Sports	Volleyball, competitive, in gymnasium
3.0	Sports	Volleyball, noncompetitive; 6–9 member team, general
8.0	Sports	Volleyball, beach
6.0	Sports	Wrestling (one match = 5 min)

7.0	Sports	Wallyball, general
2.0	Transportation	Automobile or light truck (not a semi) driving
2.0	Transportation	Flying airplane
2.5	Transportation	Motor scooter, motor cycle
6.0	Transportation	Pushing plane in and out of hangar
3.0	Transportation	Driving heavy truck, tractor, bus
7.0	Walking	Backpacking, general
3.5	Walking	Carrying infant or 15-lb load (e.g., suitcase), level ground or downstairs
9.0	Walking	Carrying load upstairs, general
5.0	Walking	Carrying 1- to 15-lb load, upstairs
6.0	Walking	16- to 24-lb load, upstairs
8.0	Walking	Carrying 25- to 49-lb load, upstairs
10.0	Walking	Carrying 50- to 74-lb load, upstairs
12.0	Walking	Carrying 74+-lb load, upstairs
7.0	Walking	Climbing hills with 0- to 9-lb load
7.5	Walking	Climbing hills with 10- to 20-lb load
8.0	Walking	Climbing hills with 21- to 42-lb load
9.0	Walking	Climbing hills with 42+-lb load
3.0	Walking	Downstairs
6.0	Walking	Hiking, cross country
6.5	Walking	Marching, rapidly, military
2.5	Walking	Pushing or pulling stroller with child
6.5	Walking	Race walking
8.0	Walking	Rock or mountain climbing
8.0	Walking	Up stairs, using or climbing up ladder
4.0	Walking	Using crutches
2.0	Walking	Walking, less than 2.0 mph, level ground, strolling, household walking, very slow
2.5	Walking	Walking, 2.0 mph, level, slow pace, firm surface
3.0	Walking	Walking, 2.5 mph, firm surface
3.0	Walking	Walking, 2.5 mph, downhill
3.5	Walking	Walking, 3.0 mph, level, moderate pace, firm surface
4.0	Walking	Walking, 3.5 mph, level, brisk, firm surface
6.0	Walking	Walking, 3.5 mph, uphill
4.0	Walking	Walking, 4.0 mph, level, firm surface, very brisk pace
4.5	Walking	Walking, 4.5 mph, level, firm surface, very, very brisk
3.5	Walking	Walking, for pleasure, work break, walking the dog
5.0	Walking	Walking, grass track
4.0	Walking	Walking, to work or class
2.5	Water activities	Boating, power
4.0	Water activities	Canoeing, on camping trip
7.0	Water activities	Canoeing, portaging
3.0	Water activities	Canoeing, rowing, 2.0–3.9 mph, light effort
7.0	Water activities	Canoeing, rowing, 4.0–5.9 mph, moderate effort
12.0	Water activities	Canoeing, rowing, >6 mph, vigorous effort
3.5	Water activities	Canoeing, rowing, for pleasure, general
12.0	Water activities	Canoeing, rowing, in competition, or crew or sculling
3.0	Water activities	Diving, springboard or platform
5.0	Water activities	Kayaking
4.0	Water activities	Paddleboat
3.0	Water activities	Sailing, boat and board sailing, wind-surfing, ice sailing, general
5.0	Water activities	Sailing, in competition
3.0	Water activities	Sailing, Sunfish/Laser/Hobby Cat, keel boats, ocean sailing, yachting
6.0	Water activities	Skiing, water
7.0	Water activities	Skimobiling
12.0	Water activities	Skindiving or scuba diving as frogman
16.0	Water activities	Skindiving, fast
12.5	Water activities	Skindiving, moderate
7.0	Water activities	Skindiving, scuba diving, general
5.0	Water activities	Snorkeling

METs	Activity category	Specific activity
3.0	Water activities	Surfing, body or board
10.0	Water activities	Swimming laps, freestyle, fast, vigorous effort
8.0	Water activities	Swimming laps, freestyle, slow, moderate or light effort
8.0	Water activities	Swimming, backstroke, general
10.0	Water activities	Swimming, breaststroke, general
11.0	Water activities	Swimming, butterfly, general
11.0	Water activities	Swimming, crawl, fast (75 yards $\cdot$ min^{-1}), vigorous effort
8.0	Water activities	Swimming, crawl, slow (50 yards $\cdot$ min^{-1}), moderate or light effort
6.0	Water activities	Swimming, lake, ocean, river
6.0	Water activities	Swimming, leisurely, not lap swimming, general
8.0	Water activities	Swimming, sidestroke, general
8.0	Water activities	Swimming, synchronized
10.0	Water activities	Swimming, treading water, fast vigorous effort
4.0	Water activities	Swimming, treading water, moderate effort, general
10.0	Water activities	Water polo
3.0	Water activities	Water volleyball
5.0	Water activities	Whitewater rafting, kayaking, or canoeing
6.0	Winter activities	Moving ice house (set up/drill holes, etc.)
5.5	Winter activities	Skating, ice, 9 mph or less
7.0	Winter activities	Skating, ice, general
9.0	Winter activities	Skating, ice, rapidly, more than 9 mph
15.0	Winter activities	Skating, speed, competitive
7.0	Winter activities	Ski jumping (climb up carrying skis)
7.0	Winter activities	Skiing, general
7.0	Winter activities	Skiing, cross-country, 2.5 mph, slow or light effort, ski walking
8.0	Winter activities	Skiing, cross-country, 4.0–4.9 mph, moderate speed and effort, general
9.0	Winter activities	Skiing, cross-country, 5.0–7.9 mph, brisk speed, vigorous effort
14.0	Winter activities	Skiing, cross-country, >8.0 mph, racing
16.5	Winter activities	Skiing, cross-country, hard snow, uphill, maximum
5.0	Winter activities	Skiing, downhill, light effort
6.0	Winter activities	Skiing, downhill, moderate effort, general
8.0	Winter activities	Skiing, downhill, vigorous effort, racing
7.0	Winter activities	Sledding, tobogganing, bobsledding, luge
8.0	Winter activities	Snow shoeing
3.5	Winter activities	Snowmobiling

Additional Values

2.0 Home activities (weaving at loom, sitting)
1.8 Music playing (accordion)
6.0 Occupation: lifting 22 lb 1 m
8.0 Occupation: lifting 45 lb 1 m
11.0 Occupation: lifting 65 lb 1 m
8.0 Walking, ice climbing
8.0 Sports: rollerskiing, 10 mph, no grade
10.0 Sports: rollerskiing, 11 mph, no grade
11.0 Sports: rollerskiing, 12 mph, no grade
12.0 Sports: rollerskiing, 9 mph, 6% grade
7.5 Sports: in-line skating, 10 mph
8.5 Sports: in-line skating, 11 mph
10.0 Sports: in-line skating, 12 mph
7.0 Water activities: underwater swimming, 1 mph
9.0 Winter activities: figure skating
14.0 Winter activities: skiing, competitive, short periods

REFERENCES

Ainsworth, B.E., Haskell, W.L., Leon, A.S., Jacobs, D.S., Jr., Montoye, H.J., Sallis, J.F., & Paffenbarger, R.S., Jr. (1993). Compendium of physical activities: Classification of energy costs of human physical activities. *Medicine and Science in Sports and Exercise*, **25**, 71-80.

Bannister, E.W., & Brown, S.R. (1968). The relative energy requirements of physical activity. In H.B. Falls (Ed.), *Exercise physiology* (pp. 267-322). New York: Academic Press.

Blair, S.N., Haskell, W.L., Ho, P., Paffenbarger, R.S., Jr., Vranizan, K.M., Farquhar, J.W., & Wood, P.D. (1985). Assessment of habitual physical activity by a 7-day recall in a community survey and controlled experiment. *American Journal of Epidemiology*, **122**, 794-804.

Collins, M.A., Cureton, K.J., Hill, D.W., & Ray, C.A. (1991). Relationship of heart rate to oxygen uptake during weight lifting exercise. *Medicine and Science in Sports and Exercise*, **23**, 636-640.

Durnin, J.V.G.A., & Passmore, R. (1967). *Energy, work and leisure*. London: Heinemann Educational Books.

Geissler, C.A., Brun, T.A., Mirbagheri, I., Soheli, A., Naghibi, A., & Hedayat, H. (1981). The energy expenditure of female carpet weavers and rural women in Iran. *American Journal of Clinical Nutrition*, **34**, 2776-2783.

Getchell, L.H. (1968). Energy cost of playing golf. *Archives of Physical Medical Rehabilitation*, **49**, 31-35.

Goff, L.G., Frassetto, R., & Specht, H. (1956). Oxygen requirement in underwater swimming. *Journal of Applied Physiology*, **9**, 219-221.

Howley, E.T., & Glover, M.E. (1974). The caloric costs of running and walking one mile for men and women. *Medicine and Science in Sports and Exercise*, **6**, 235-237.

Leon, A.S. (1981). Approximate energy expenditures and fitness values of sports and recreational and household activities. In E.L. Wynder (Ed.), *The book of health and physical fitness* (pp. 283-341). New York: Watts.

Mandli, M., Hoffman, M.D., Jones, G.M., Bota, B., & Clifford, P.S. (1989). Nordic ski training: In-line skates or roller skis? *Medicine and Science in Sports and Exercise*, **21**, S64.

McArdle, W.D., Katch, F.I., & Katch, V.L. (1988). *Exercise physiology: Energy, nutrition, and human performance* (2nd ed.). Philadelphia: Lea & Febiger.

Montoye, H.J. (1982). Age and oxygen utilization during submaximal treadmill exercise in males. *Journal of Gerontology*, **37**, 396-402.

Nelson, D.J., Pells, A.E. III, Geenen, D.L., & White, T.P. (1988). Cardiac frequency and caloric cost of aerobic dancing in young women. *Research Quarterly for Exercise and Sport*, **59**, 229-233.

Passmore, R., & Durnin, J.V.G.A. (1955). Human energy expenditure. *Physiological Reviews*, **35**, 801-840.

Reiff, G.G., Montoye, H.J., Remington, R.D., Napier, J.A., Metzner, H.L., & Epstein, F.H. (1967a). Assessment of physical activity by questionnaire and interview. *Journal of of Sports Medicine and Physical Fitness*, **7**, 1-32.

Reiff, G.G., Montoye, H.J., Remington, R.D., Napier, J.A., Metzner, H.L., & Epstein, F.H. (1967b). Assessment of physical activity by questionnaire and interview. In M.J. Karvonen & A.J. Barry (Eds.), *Physical activity and the heart* (pp. 336-371). Springfield, IL: Charles C Thomas.

Seliger, V. (1968). Energy metabolism in selected physical exercises. *Internationale Zeitschrift für Angewande Physiologie Einschliesslich Arbeitsphysiologie*, **25**, 104-120.

Stray-Gundersen, J., & Galanes, J. (1991). The metabolic demands of roller skating compared to snow skiing [abstract]. *Medicine and Science in Sports and Exercise*, **23**, S107.

Torún, B. (1983). Inaccuracy of applying energy expenditure rates of adults to children. *Journal of Clinical Nutrition*, **38**, 813-814.

Torún, B., Chew, F., & Mendoza, R.D. (1983). Energy cost of activities of preschool children. *Nutrition Research*, **3**, 401-406.

Veicsteinas, A. Ferretti, G., Margonato, V., Rosa, G., & Tagliabue, D. (1984). Energy cost of and energy sources for alpine skiing in top athletes. *Journal of Applied Physiology*, **56**, 1187-1190.

VonHofen, D., Auble, T.E., & Schwartz, L. (1989). Aerobic requirements for pumping versus carrying .9 kg hand weights while running [abstract]. *Medicine and Science in Sports and Exercise*, **21**, S7.

Watts, P.B., Martin, D.T., Schmeling, M.H., Silta, B.C., & Watts, A.G. (1990). Exertional intensities and energy requirements of technical mountaineering at moderate altitudes. *Journal of Sports Medicine and Physical Fitness*, **30**, 365-376.

Wigaeus, E., & Kilbom, A. (1980). Physical demands during folk dancing. *European Journal of Applied Physiology*, **45**, 117-183.

Appendix D

Tecumseh and Minnesota Occupational and Leisure Time Activity Questionnaires

Tecumseh Self-Administered Occupational Activity Questionnaire

This questionnaire is designed to be self-administered. However, when the questionnaire is picked up, the interviewer should check it over very carefully to see that there were no omissions, that the occupation(s) are clearly understood, that the answers seem reasonable, and that the additional tasks (Question 27) are clearly understood by the interviewer, because metabolic rates will have to be designated for these activities.

Section A

Layoffs should be reflected by reduced months employed in Question 2 in this section. In Questions 4 and 5, occupation and name of business are for reference only. They will not be coded.

Section B

In Question 7, count all months in which respondent worked at least 2 weeks. In Question 8, for some persons transportation to and from work will be the only source of physical activity. Make certain the distances are recorded as round-trip distances. The respondent may indicate that he or she travels from one job to another without going home. In these instances, if the transportation is by auto, bus, or subway, ignore the information. If the transportation is by walking or biking, ask sufficient questions to be able to calculate and record the actual distances traveled. Include walking distance from parking lot or bus stop to job site.

Section C

Question 11:

Frequently truck drivers load and unload cargo. Make certain this is reflected in the responses. If there is essentially no loading or unloading, only #11 should be completed insofar as driving a truck is concerned. If there is loading and unloading, this should be reflected in questions 19, 20, 21, 22, or 23.

Question 25 refers to use of heavy power tools of various sorts.

Question 26 involves no power equipment.

Section D

Question 29

There is more difficulty estimating time spent ascending or descending stairs, so the respondent is asked to indicate the approximate number of flights per day. Make certain he or she indicates this rather than hours. One flight is equivalent to about 10 steps and includes ascending and descending as one flight. Include only work days.

Special note for use of the Minnesota Leisure Time Physical Activity Questionnaire (LTPA):

Reprinted from the *Journal of Chronic Diseases*, **31**. A questionnaire for the assessment of leisure time physical activities, Taylor, H.L., Jacobs, D.R., Schucker, B., Knudsen, J., Leon, A.S., & DeBacker, G., pages 741-755, copyright 1978, with kind permission from Pergamon Press Ltd, Headington Hill Hall, Oxford 0X3 0BW, UK.

The final table, "Additional Activities and Intensity Codes," is reprinted with permission from Folsom, A.R., Caspersen, C.J., Taylor, H.J., Jacobs, D.R., Jr., Luepker, R.V., Gomez-Marin, O., Gillum, R.F., & Balckburn, H. (1985). Leisure time physical activity and its relationship to coronray risk factors in a population-based sample: The Minnesota heart survey. *American Journal of Epidemiology*, **121**, 570-579. Baltimore, MD: American Journal of Epidemiology.

Section A Work History

Please fill in the following information about jobs you have held in the past 12 months. Jobs include both full-time and part-time. You may have a single job or hold two or more jobs at once. If you change responsibilities with the same employer, consider it to be a change of jobs if a substantial change occurs in physical effort or tasks demanding physical activity. For example, changing from bookkeeping to construction work within the same company would be a "change of job." If you have held more than three jobs in the past 12 months, please answer for the three jobs that required the most physical effort.

1. Have you been employed at a job for pay in the past 12 months? 1. Yes ____ → Go to Question 2; 2. No ____ → You are done with this questionnaire

2. For how many months were you employed in the past 12 months? __ __ months
3. How many weeks were you on vacation (not working in any job)? __ __ weeks

Work history for the past 12 months

	Job 1	Job 2	Job 3
4. Occupation:	________	________	________
5. Name of business:	________	________	________
6. How many hours/week do/did you usually work?	________	________	________

Section B Transportation to and From Work

7. Check the months you worked each job. Then check the months you used each kind of transportation to and from each of your jobs.

	Job 1											
	Jan	Feb	Mar	Apr	May	Jun	Jul	Aug	Sep	Oct	Nov	Dec
Months worked each job:												
Mode of transportation	Job 1											
Auto, bus or subway												
Bicycle												
Walking												

	Job 2											
	Jan	Feb	Mar	Apr	May	Jun	Jul	Aug	Sep	Oct	Nov	Dec
Months worked each job:												
Mode of transportation	Job 2											
Auto, bus or subway												
Bicycle												
Walking												

	Job 3											
	Jan	Feb	Mar	Apr	May	Jun	Jul	Aug	Sep	Oct	Nov	Dec
Months worked each job:												
Mode of transportation	Job 3											
Auto, bus or subway												
Bicycle												
Walking												

Continued

8. What is the approximate *round-trip* distance that you walked or bicycled to and from each of your jobs *in miles or blocks*?

(If you never walked or bicycled, enter "0" for distance. Do *not* include walking or bicycling on the job.)

	Job 1	Job 2	Job 3
Distance walked:	miles ____ blocks ____	miles ____ blocks ____	miles ____ blocks ____
Distance bicycled:	miles ____ blocks ____	miles ____ blocks ____	miles ____ blocks ____

9. In months that you *walked* to work, how many times per month did you do this?

	Job 1	Job 2	Job 3
Average times per month *walked* to work:	____	____	____

10. In months that you *bicycled* to work, how many times per month did you usually do this?

	Job 1	Job 2	Job 3
Average times per month *bicycled* to work:	____	____	____

Section C. Physical Activity at Work

We are interested in the amount of activity that people do in different occupations. Think about job-related activities you mostly did in a work week during the time you worked in the past 12 months. First, read each of the following categories and mark "Yes" for the ones you usually did and "No" for the ones you did not do. Then go back through the list and record the number of *hours per week* you spent on each activity.

Please note: The number of hours in each activity should add up to the number of hours you worked in each job, as you stated in Section A, question 6.

	Job 1		Job 2		Job 3	
		hr/wk		hr/wk		hr/wk
11. *Sitting—light work* *Desk work *Using hand tools *Light assembly or repair *Driving car or truck (if you carry or move heavy objects, respond to Items 19-23 also)	1. ____ Yes → 2. ____ No	____	1. ____ Yes → 2. ____ No	____	1. ____ Yes → 2. ____ No	____
12. *Sitting—moderate work* *Working heavy levers Riding mower or forklift *Crane operation	1. ____ Yes → 2. ____ No	____	1. ____ Yes → 2. ____ No	____	1. ____ Yes → 2. ____ No	____

Continued

	Job 1	hr/wk	Job 2	hr/wk	Job 3	hr/wk
13. Standing—light work *Bartender/lab tech work *Store clerk behind a counter *Standing talking with co-worker or customer/filing papers *Assembling light machine parts at own pace *Using hand tools/car wash attendant	1. ____ Yes → 2. ____ No	____	1. ____ Yes → 2. ____ No	____	1. ____ Yes → 2. ____ No	____
14. Standing—light to moderate work *Light welding/stocking shelves *Assembling or repairing heavy machine parts *Packing or unpacking small objects *Sanding, scrubbing, polishing floors with power equipment	1. ____ Yes → 2. ____ No	____	1. ____ Yes → 2. ____ No	____	1. ____ Yes → 2. ____ No	____
15. Standing—moderate work *Pulling on wires/twisting cables *Assembling or work with light machine parts at fast rate on assembly line *Lifting up to 50 lb every 5 min or so for a few seconds at a time *Cranking up dollies/hitching trailers *Operating large levers, jacks *Jerking on ropes, cables, etc. (such as rewiring houses)	1. ____ Yes → 2. ____ No	____	1. ____ Yes → 2. ____ No	____	2. ____ Yes → 2. ____ No	____
16. Standing—moderate to heavy work *Masonry/painting *Lifting more than 50 lb every 5 min or so for a few seconds at a time	1. ____ Yes → 2. ____ No	____	1. ____ Yes → 2. ____ No	____	1. ____ Yes → 2. ____ No	____
17. Walking at work, not carrying anything heavier than briefcase *Between buildings/in hallways *Roving store clerk	1. ____ Yes → 2. ____ No	____	1. ____ Yes → 2. ____ No	____	1. ____ Yes → 2. ____ No	____
18. Walking, carrying something *Trays, dishes *Gas station mechanic, changing tires, wrecker work, etc.	1. ____ Yes → 2. ____ No	____	1. ____ Yes → 2. ____ No	____	1. ____ Yes → 2. ____ No	____
19. Standing or walking, carrying objects about 25 lb	1. ____ Yes → 2. ____ No	____	1. ____ Yes → 2. ____ No	____	1. ____ Yes → 2. ____ No	____

	Job 1		Job 2		Job 3	
		hr/wk		hr/wk		hr/wk
20. Standing or walking, carrying objects about 50 lb	1. ____ Yes → 2. ____ No	____	1. ____ Yes → 2. ____ No	____	1. ____ Yes → 2. ____ No	____
21. Standing or walking, carrying objects about 75 lb	1. ____ Yes → 2. ____ No	____	1. ____ Yes → 2. ____ No	____	1. ____ Yes → 2. ____ No	____
22. Standing or walking, carrying objects about 100 lb	1. ____ Yes → 2. ____ No	____	1. ____ Yes → 2. ____ No	____	1. ____ Yes → 2. ____ No	____
23. Moving, pushing heavy objects, 75 lb or more	1. ____ Yes → 2. ____ No	____	1. ____ Yes → 2. ____ No	____	1. ____ Yes → 2. ____ No	____
24. Carpentry	1. ____ Yes → 2. ____ No	____	1. ____ Yes → 2. ____ No	____	1. ____ Yes → 2. ____ No	____
25. Using heavy power tools, such as pneumatic tools (drills, jackhammers, tampers)	1. ____ Yes → 2. ____ No	____	1. ____ Yes → 2. ____ No	____	1. ____ Yes → 2. ____ No	____
26. Using heavy tools (not power tools), such as shovel, pick, tunnel bar, spade	1. ____ Yes → 2. ____ No	____	1. ____ Yes → 2. ____ No	____	1. ____ Yes → 2. ____ No	____
27. Other activities (specify)						
________________	1. ____ Yes →	____	1. ____ Yes →	____	1. ____ Yes →	____
________________	2. ____ No		2. ____ No		2. ____ No	
28. Total the number of hours for each job. This should equal the totals in Question 6.		____		____		____

Section D Walking Up and Down Stairs or Ladders

29. Giver the number of flights of stairs or steps per day. Use 10 stairs per flight, and count up and down as one flight.	____	____	____

*Modified from Montoye, H.J. (1989). Lessons from Tecumseh on the assessment of physical activity and fitness. In T.F. Daury (Ed.), *Assessing Physical Fitness and Physical Activity in Population-based Surveys.* U.S. Government Printing Office, DHHS Publication No. (PHS) 89-1253, pp. 349-376.

Due to overlap, the Minnesota LTPA must be modified slightly for use with this questionnaire. Specifically, Minnesota LTPA item 020, "walking to work," should not be answered; Minnesota LTPA item 115, "biking for pleasure or to work" should be modified to "biking for pleasure;" Minnesota LTPA item 030, "using stairs when elevator is available," should be answered onlyfor leisure time. Intensity codes for these activities remain unchanged.

Tecumseh Occupation Physical Activity MET Values

Use the following MET values for transportation activities:

Walking	3.5 METs
Bicycling	4.0 METs
Automobile, bus, train	1.5 METs

Use these METs for questions 11 through 26 regarding job activities.

Question	*METs*
11	1.5
12	2.5
13	2.5
14	3.0
15	3.5
16	4.0
17	3.5
18	4.5
19	4.0
20	6.5
21	7.5
22	8.5
23	4.5
24	6.0
25	8.0
26	8.0

The Minnesota Leisure Time Physical Activity Questionnaire

The administration of the physical activity questionnaire requires that special attention be paid to interviewing technique to limit bias in the data and to prevent the interview from becoming ponderous and irritating for both the participant and the interviewer. It is difficult for most people to remember what they did the previous year, especially when it comes to an activity such as walking. Some participants tend to give up and do not try to make an estimate. Other participants take the task very seriously and try too hard, dragging out the interview unnecessarily.

As an interviewer you should establish rapport during the introduction, perhaps exchanging a few pleasantries. Stressing the importance of the data can be achieved by emphasizing key words. Instructions should be given in a slow, clear manner. From this point on you should take the initiative and set and maintain the pace in a very matter-of-fact way. Extraneous talk should be avoided. Though a participant should not be hurried, if he is spending undue time trying to recall detail, you should interrupt with "Remember we're interested in an average or an estimate, not an exact time" or "In general what would you say?"

If a participant rambles you should politely cut in with a reminder that you are interested in months and average times.

For any participant, challenge anything that seems exaggerated.

Example: A participant states that he swims one hour a week at the YMCA. Make sure that hour does not include changing time and socializing. In fact, the actual swimming time may be only 20 min.

If a participant states that he performs an activity more than 8 times a month (which would average twice a week), translate it into weeks to verify it.

Example: Participant states he plays softball 16 times a month. The interviewer should state "on the average you play softball 4 times a week."

If a participant says he does something during the summer months, do not assume which months he means. Probe "Which months are you referring to?"

It is known that people tend to overestimate time spent at a particular activity. If a participant says "2 or 3 hours," record 2. If the range given is large ("5 to 10 times"), ask that he try to be more specific.

If an activity is performed very frequently, "number of times/month" may be a difficult time reference. Suggest that the participant think in terms of number of times per week.

If an outdoor activity is performed every month, probe "Is your activity the same in summer as in winter?"

Expect that an average interview will last 10 to 20 min. Your goal is to get estimates as accurate as possible while maintaining a moderate to brisk, interesting pace. The more experience you get interviewing, the easier these techniques will be. Hearing tapes of your own interviews would be extremely helpful in developing style. Plan to interview and record at least six respondents (staff members are fine) as practice. Evaluate your style and how situations might have been handled differently—perhaps more information needed, an unchallenged questionable response, a little faster pace, etc.

Sample Introduction

Science does not yet have all the answers regarding the relationship between physical activity and coronary heart disease. It is *very important* that we collect physical activity data for each participant. We use this form and will ask you to make the *best estimate* you can in answering the questions.

Sample Instruction

In this column we have listed different kinds of physical activities (*point*). In this column you checked whether you did or did not perform an activity in the last year. Is that right? So these activities with a check in the "Yes" column (*point*) are activities you performed sometime between now and last June (*appropriate month*). For each of these activities, I'd like you to tell me in which months you performed them (*point*), then I'd like the average number of times per month (*point*), and lastly the average time you spent at the activity each time you performed it.

For example, suppose you had checked backpacking. First you would give the months. Let's say July and August. Then you would tell me the average *number* of times per month. If it was once in July and three times in August, you would tell me twice. Then you would tell me the average *time* you spent at the activity *each* time you backpacked.

This may sound confusing, but once we start a routine it will be quite easy. (A positive attitude and manner here are very important!)

For the first one or two activities checked, you will have to go through the steps verbatim.

Example: "You've checked that you've done backpacking. In which months did you backpack? What was the average number of times each month?"

After this you should strive for word economy and use just words or phrases rather than entire sentences. Pointing with a pen helps.

Example: "Sailing. Months? Average times per month? Time per occasion?" Once the routine is well established, after starting the activity, a nod of the head and pointing with a pen should elicit a response.

Activities

Several activities will require special probing or clarifying comments for *each* participant, no matter how the first column has been checked. A definition has been written for each of the activities. You should be familiar with all of these. With the exception of the categories mentioned below, you need not define activities unless a participant has a question.

Four situations in which walking is done from point to point requiring *continuous* walking for 10 min or more are requested as separate items. Note that time of walking during working hours is not wanted except for long breaks, such as lunch.

For using stairs instead of elevator, state "For this *one* activity we will consider your choice of stairs over an elevator at any time, even during your work day." Then probe in the routine manner.

If a participant has checked home exercise or health club, ask what the specific activities are and record data per instructions in the "Definitions" section.

Under home repair activities, state to *each* participant, "Because of space limitations, we couldn't list all possible home activities. Can you think of any other major repair or maintenance job which you did last year?" (Note: Under the definitions section are procedures for coding the home repair category.)

The lawn-mowing categories require some clarification. If a participant checks *walking behind power mower*, state "By this we mean a power mower which has to be pushed or a self-propelled mower." If *pushing hand mower* is checked, state "By this we mean a mower which has no power."

The following activities have constraints (see "Definitions"). If any of these activities is checked, probe to be sure the constraint has been met:

Swimming
Cross-country hiking
Backpacking
Bicycling
Sailing

Standardized Times for Activities

To insure uniformity we will consider the following:

4 weeks in a month
48 work weeks per year
240 work days per year
22 work days per month
100 weekend days per year

Standard times have been established for the following activities:

1 flight stairs = 1/2 min (round up to the nearest minute)
1 water ski ride = 5 min
1 bowling game or line = 10 min
1 tennis singles set = 20 min
1 tennis doubles set = 30 min

Name ______________________

Leisure Time Physical Activities

Listed below are a series of leisure time activities. Related activities are grouped under general headings. Please read the list and check "Yes" in Column 4 for those activities which you have performed in the last 12 months, and "No" in Column 3 for those you have not. Do not complete any of the other columns.

To be completed by participant **Activity (1)**		Did you perform this activity?		DO NOT WRITE IN THIS SPACE	Month of activity												Average number of times per month	Times per occassion	
		No	Yes		Jan	Feb	Mar	Apr	May	Jun	July	Aug	Sep	Oct	Nov	Dec		Hr	Min
Section A: Walking and miscellaneous																			
010	Walking for pleasure																		
020	Walking to work																		
030	Using stairs when elevator is available																		
040	Cross-country hiking																		
050	Backpacking																		
060	Mountain climbing																		
115	Bicycling for pleasure																		
125	Dancing—ballroom, square, and/or disco																		
135	Dancing—aerobic, ballet																		
140	Horseback riding																		
Section B: Conditioning exercise																			
150	Home exercise																		
160	Health club exercise																		
180	Jog/walk combination																		
200	Running																		
210	Weight lifting																		

To be completed by participant **Activity (1)**		Did you perform this activity?		DO NOT WRITE IN THIS SPACE	Month of Activity												Average number of times per month	Times per occassion	
					Jan	Feb	Mar	Apr	May	Jun	July	Aug	Sep	Oct	Nov	Dec		Hrs	Min
Section C: Water activities		No	Yes																
220	Water skiing																		
235	Sailing in competition																		
250	Canoeing or rowing for pleasure																		
260	Canoeing or rowing in competition																		
270	Canoeing on a camping trip																		
280	Swimming (at least 50 ft) at a pool																		
295	Swimming at the beach																		
310	Scuba diving																		
320	Snorkeling																		
Section D: Winter activities																			
340	Snow skiing, downhill																		
350	Snow skiing, cross-country																		
360	Ice or roller skating																		
370	Sledding or tobogganing																		
Section E: Sports																			
390	Bowling																		
400	Volleyball																		
410	Table tennis																		
420	Tennis, singles																		
430	Tennis, doubles																		

	To be completed by participant **Activity (1)**	Did you perform this activity?		DO NOT WRITE IN THIS SPACE	Month of Activity												Average number of times per month	Times per occassion	
		No	Yes		Jan	Feb	Mar	Apr	May	Jun	July	Aug	Sep	Oct	Nov	Dec		Hrs	Min
	Section E: Sports (continued)																		
440	Softball																		
450	Badminton																		
460	Paddleball																		
470	Racquetball																		
480	Basketball: nongame																		
490	Basketball: game play																		
500	Basketball: officiating																		
510	Touch football																		
520	Handball																		
530	Squash																		
540	Soccer																		
	Golf:																		
070	Riding a power cart																		
080	Walking, pulling clubs on cart																		
090	Walking and carrying clubs																		

	To be completed by participant **Activity (1)**	Did you perform this activity? No	Yes	DO NOT WRITE IN THIS SPACE	Month of Activity Jan	Feb	Mar	Apr	May	Jun	July	Aug	Sep	Oct	Nov	Dec	Average number of times per month	Times per occassion Hrs	Min
	Section F: Lawn and garden activities																		
550	Mowing lawn with riding mower																		
560	Mowing lawn walking behind power mower																		
570	Mowing lawn pushing hand mower																		
580	Weeding and cultivating garden																		
590	Spading, digging, filling in garden																		
600	Raking lawn																		
610	Snow shoveling by hand																		
	Section G: Home repair activities																		
620	Carpentry in workshop																		
630	Painting inside of house, includes paper hanging																		
640	Carpentry outside																		
650	Painting outside of house																		
	Section H: Fishing and hunting																		
660	Fishing from riverbank																		
670	Fishing in stream with wading boots																		
680	Hunting pheasants or grouse																		
690	Hunting rabbits, prairie chickens, squirrels, raccoon																		
710	Hunting large game: deer, elk, bear																		
	Section I: Other activities																		

Table D.1 Sample Worksheet for Tecumseh Occupational and Minnesota Leisure Time Physical Activity Questionnaires Combined

00621 — ID number

John Doe — Name

Activity	METs	Hr/week (mean for year)	METs × Hr/week
Occupation			
Sitting	1.5	18.0	27.00
Standing (light work)	2.5	22.0	55.00
Walking (carrying about 30 lb)	4.0	8.0	32.00
Active leisure			
Rowing (pleasure)	3.0	0.2	0.60
Boat fishing	2.5	0.9	2.25
Hunting (general)	5.0	0.4	2.00
Transportation to work			
Auto	1.5	1.5	2.25
Bicycle	4.0	2.2	8.80
Meals[a]	1.8	21.0	37.80
Sleep[a]	1.0	56.0	56.00
Total hours		127.0	
Quiet leisure[b]			
Reading, watching TV	1.5	41.0	61.50
Total		168	285.20
Mean METs/day 285.2 ÷ 168 = 1.70[c]			

[a]Unless other information is available, it is assumed each subject averages 8 hr per day sleeping and 3 hr per day eating.
[b]The number of hours engaged in quiet leisure is obtained by subtracting the number of hours engaged in other activities (127 in this case) from 168, the total number of hours in a week.
[c]This figure may be converted to kilocalories per 24 hr by multiplying it by 24 and by the subject's weight in kilograms. The various activities may be coded and a computer program written to do all the calculations on this worksheet.

Softball, 7 innings: pitcher or catcher = 20 min; other players = 10 min
Softball, 9 innings: pitcher or catcher = 25 min; other players = 15 min
9 holes of golf = 1 hr, 30 min

For these activities the questioning format will change, i.e., do not ask how many minutes you spend on the stairs, ask the number of flights and translate into minutes using standard of 1/2 min per flight, or how many holes of golf, etc.

Miscellaneous

If the participant has filled out the entire questionnaire, checking months, etc., it will be necessary to validate each activity by reading back to him the information he has put down. Question anything which looks out of the ordinary.

Definitions and Comments on Activities

Code Title and comment (METs)

Walking

010 ***Walking for pleasure***. Since this is the most frequently reported activity, each man should be asked specifically about it. (3.5)

015 ***Walking to and from work***. Walking from the bus to work, or the parking lot, etc., may be included in this category if the walking is continuous for 10 or more minutes. Such walking may be repeated in the evening. (4.0)

020 ***Walking during work breaks***. Include only walking which is not connected with work, such as walks during lunch hour. Walking which is associated with customary performance of the occupation is not included. (3.5)

030 ***Voluntarily using stairs when elevator or escalator is available***. Ask specifically for the number of trips upstairs. Do not count walking downstairs: count 1/2 min for each flight (a flight = one story). (8.0)

040 ***Cross-country hiking***. Walking continuously on flat or in hilly terrains without backpack for at least 2 hr. Ask for elapsed time and frequency and duration of rest periods and stops for eating. Finally estimate total time spent walking per occasion. (6.0)

050 ***Backpacking***. Defined as walking and carrying a pack weighing 20 lb or more containing for instance, gear and supplies for overnight camping. If the activity does not qualify, record time, etc., under 040. Discount stops for rest and eating, etc. (7.0)

060 ***Mountain climbing***. Walking trips in which the purpose is to reach a 'high point' which takes several hours or days to accomplish qualify as mountain climbing. No distinction between rock climbing or hill climbing is made. Ask subject to distinguish between actual climbing and rest stops, eating, sleeping, etc. Total time to include both up and down time. (8.0)

115 ***Bicycling***. To work and/or for pleasure. No distinction is made regarding the type of bicycle or the terrain. Ask for actual riding time. The rare individual who engages in medium and long distance racing should be reported under Other Activities under the title of *Competitive Bicycling*. Obtain data on practice sessions, include races as practice sessions. (4.0)

125 ***Dancing***. Ballroom and/or square dancing. Ask for time spent on dance floor. (5.5)

Conditioning Exercises

Setting up exercise, special routines for increasing flexibility or strength, running in place for roughly 3 min, carried out at home or in the health club should be reported under 'Home exercise' or 'Health club.'

On the other hand, if a participant concentrates his activities in one area such as jog-walk, jogging, running, weight lifting, do not report this under 'Home exercise' or 'Health club,' but under the activity (180, 200, 210). If a participant goes to a YMCA for the sole purpose of playing a game of squash or other games listed in the section headed *Sports*, please list this activity under the specific game rather than under activities at 'Health club.'

If the company which employs the participant offers physical conditioning facilities and exercise routines on company time, these should be reported under 'Health club' or the particular activities listed above.

150 ***Home exercise***. Ask what kind of exercise is done. Do not include items listed under other codes. Ask for *time spent actually exercising*. (4.5)

160 ***Health club***. Ask what kind of exercise is done at the club. Distinguish between visits for exercise classes and visits to engage in a single specific game (such as volleyball) or specific activity (such as swimming). Report specific activities below. Ask for total time spent in locker room, steam room, etc. (6.0)

180 ***Jogging-walking***. Ask for time spent in jogging and walking (most participants who do this will have a good estimate of the time). (6.0)

200 ***Running***. Ask for time spent running. 'Running' is defined as continuous running for at least 10 min, using full length strides. Shorter continuous activity is to be reported under jog-walk. (8.0-12.0)

210 ***Weight lifting***. Ask for time spent in the weight-lifting area. The type of weight lifting is not important for the purposes of this question. (3.0-6.0)

Water Activities

220 ***Water skiing***. To obtain *time per occasion* ask for the total number of 'rides' per occasion. Multiply the number of rides or trips by 5 and record this as the total minutes of activity. (6.0)

235 ***Sailing***. Only those individuals who sail in racing competition are to be recorded here. Record the number of hours per occasion the participant is either racing or practicing. (3.0)

250 ***Canoeing or rowing for pleasure***. Record the hours per occasion. Be sure the participant distinguishes riding in row boat from rowing. (3.5)

260 ***Canoeing or rowing in competition***. Ask for the number of months of training. Number of training sessions per month. Average time per training session. (12.0)

270 ***Canoeing on a camping trip***. Include the time paddling, whether bow or stern. Also included is associated activities such as portage, setting up camp, and maintaining camp. (4.0)

280 ***Swimming (at a pool)***. Distinguish between time spent sitting in the sun by the side of the motel pool (drinking beer or bloody Marys?), time spent 'cooling off,' and time spent actually swimming. Was the pool large enough to swim in? Athletic clubs and YMCAs have pools 50–75 ft in length. Verify. (6.0)

295 ***Swimming at the beach or lake.*** Time spent in sitting on the beach or playing with waterball in 18 in of water is not wanted. Ask if participant swam out into deep water and how long was he swimming in water over his head? Do not include time spent snorkeling. (6.0)

310 ***Scuba diving.*** Swimming under water while breathing oxygen from a tank strapped to back. Ask for time actually swimming under water. (7.0)

320 ***Snorkeling.*** Swimming with a face mask and breathing tube. Ask for time in water with snorkel gear in place. (5.0)

Winter Activities

340 ***Snow skiing downhill.*** Ask for time spent actually skiing downhill. It may help to ask the participant to estimate the number of runs per occasion and roughly how long each run actually took. Competitive downhill racing should be reported under 'Other activities' with the title Downhill Ski Racing. (7.0)

350 ***Cross-country skiing.*** Ask the subject for the average amount of time spent cross-country skiing. If a respondent reports Snow Shoeing, that can also be recorded in this section of the form. (8.0)

360 ***Ice or roller skating.*** Total time spent at rink, minus rest periods and socializing. (7.0)

370 ***Sledding or tobogganing.*** As for time on the slope. Then how much time is spent walking uphill. Report time walking uphill. If mechanical transportation is provided for going back up the hill, do not report activity. (7.0)

Sports

390 ***Bowling.*** Ask the participant "How many games or lines do you bowl on an average night or occasion?" Mulitply the number of games times 10. The answer is time per occasion in minutes. (3.0)

400 ***Volleyball.*** Ask for and record time spent on court. (4.0)

410 ***Table tennis.*** Ask for and record total time playing. (4.0)

420 ***Tennis singles.*** Ask for number of sets. Multiply the number of sets by 20 min. The answer is playing time in minutes. For lessons and volleying, record court time. (6.0-8.0)

430 ***Tennis doubles.*** Ask for number of sets. Multiply the number of sets by 15 min. Record the answer which is playing time in minutes. (6.0)

440 ***Softball.*** Record the number of games. Ask for innings per game. 7 inning games are considered to last 1 hr. 30 min. 9 inning games 2 hr. (5.0)

450 ***Badminton.*** Record court time. Report tournament play in Section I under the heading of Competitive Badminton activities. (7.0)

460 ***Paddleball.*** Record court time. (6.0-8.0)

470 ***Racketball.*** Record court time. (7.0-10.0)

480 ***Basketball.*** Record court time. (6.0)

490 ***Basketball game play.*** Record court time. (8.0)

500 ***Basketball officiating.*** Record court time. (7.0)

510 ***Touch football.*** Record time of game. (8.0)

520 ***Handball.*** Record court time. (10.0-12.0)

530 ***Squash.*** Record court time. (10.0-12.0)

540 ***Soccer.*** Record total playing time. (7.0-12.0)

Golf

Identify the method of carrying clubs. Players who employ a caddy can be reported under code 80. Ask for the number of holes played. Count 1 1/2 hr for every nine holes played.

070 Riding a power cart. (3.5)

080 Walking, pulling clubs in cart. (5.0)

090 Walking and carrying clubs. (5.5)

Lawn and Garden Activities

550 ***Mowing lawn. Riding a power mower.*** Ask for average time to cut lawn. Inquire regarding coffee, coke (or beer?), or rest breaks. Adjust time accordingly. (2.5)

560 ***Mowing lawn. Walking behind a power mower.*** This classification includes mowers with power applied to cutting blades only and also includes mowers with power applied to wheels and cutting blades. Record time to cut lawn with due regard to rest time. (4.5)

570 ***Mowing lawn. Pushing hand mower.*** Record time with due regard to rest time. (6.0)

580 ***Weeding and cultivation of garden.*** This item includes all activities needed to maintain an already planted garden. It can be done several times over the gardening season. Ask the subject to estimate the amount of time it takes with due regard for rest breaks. (4.5)

590 ***Spading, digging, filling in garden.*** This item refers to the activities needed to prepare a garden for planting. It is usually one only in the spring, and so should not be checked for

consecutive months. Ask the subject to estimate the time needed with due regard for rest time. (5.0)

600 ***Raking lawn***. Record the time spent raking with due regard for rest time. (4.0)

610 ***Snow shoveling by hand***. By checking with the local office of the National Weather Bureau, the Minnesota Center established a snow shoveling rate of 4–6 times per month (for the winter of 1976-77). The criteria used for snow that required shoveling was at least 1 in. snowfall within 48 hr. If the subject gives an estimate beyond the 6 times a month limit, he is questioned further to make sure that estimate is valid. It is suggested that each center where snow shoveling would be reported check with their branch of the Weather Bureau and determine a standard to be used. (6.0)

Home Repair Activities

This section uses a limited number of specified activities to cover a large cluster of related activities. In the definition of activities given below, various other home repair activities are listed in addition to the principal heading. A good many home repair projects require from a half to several days of work. Such activities may be confined to a relatively short portion of the year. To simplify recording it is proposed that all such activities be recorded in one month, usually the vacation period. Days to complete the task should be cumulated, using 8 hr for one day. For example, Mr. Jones rebuilt a porch during his vacation (August) in 8 half-days of work and then built a brick wall spending 3–5 hr on Saturdays for 10 weekends (September-November). To compute number of days in these activities, assume 4 hr per occasion in building the brick wall. There are then 18 half-days which may be recorded under code 640 as nine occasions in August with 8 hr per occasion.

620 ***Carpentry in work shop***. Construction of furniture or comparable objects using hand held or power tools or repair of storm windows or screens or minor repairs inside the house can be included under this code. Record cumulative time spent in shop or doing minor repairs. (3.0)

630 ***Painting inside of house or wallpaper hanging***. Waxing floors, laying tile, installing or repairing plumbing, installing or repairing interior electric lines may be included under this code. Record time spent performing the task. (4.5)

640 ***Carpentry outside of house***. Building porches, garages, car ports, fences, etc., laying brick on walls or patios may be included under this code. Report cumulative time necessary to finish the job or jobs. (6.0)

650 ***Painting outside of home***. Painting outside of house, jobs which require ladders, changing storm windows and screens, washing windows, mixing and pouring cement, laying cement blocks and digging trenches for foundations may be included under this code. Report cumulative time necessary to complete one or more tasks. (5.0)

Fishing and Hunting

660 ***Fishing from river bank***. Record time (hours) spent on river bank. (3.5)

670 ***Fishing in stream with wading boots***. Record time (hours) spent actually fishing. (6.0)

680 ***Hunting pheasants or grouse***. Ask for time (hours) walking through cornfields (for pheasants) or through the woods (for grouse): pool time for both activities. (6.0)

690 ***Hunting rabbits, prairie chickens, squirrels and raccoon***. Ask for time (hours) spent in the field looking for game. (5.0)

710 ***Hunting large game—deer, elk, bear***. Record days spent in the field. (6.0)

Other Activities

There will be the occasional individual who has spent a large amount of time on an activity that is not referred to here. If this time adds up to 8 hr during the year, record under other activities, asking the participant to give a name describing this activity.

Additional Activities and Intensity Codes

Code	Activity	METs	Code	Activity	METs
135	Dancing, aerobics/ballet	6.0	380	Showshoeing	8.0
140	Horseback riding	3.5	700	Hunting ducks	2.5
170	Brisk walking	4.5	720	Hunting, bow and arrow	3.5
190	Jogging	7.0	730	Pistol shooting	2.5
330	Surfing	6.0	740	Archery	3.0

Appendix E

Paffenbarger/Harvard Alumni Questionnaire

Climbing up and down one flight daily was rated 28 kcal per week; walking one block daily, 56 kcal per week. Based on published literature (12), each sport was scored according to intensity as a ratio of work metabolic rate to resting metabolic rate (MET score). A list of the various sports ordered by their MET scores is given here. Sports with a score of ≤4 METs were deemed light; those scoring 4.5-5.5 METs, moderate; and those scoring ≥6 METs, vigorous (13). Since resting metabolic rate is approximately one kilocalorie per kilogram of body weight per hour, the energy expended per week on each sport was estimated by multiplying its MET score by weight in kilograms and weekly hours of play. An index of total energy expenditure per week was estimated by summing the total kilocalories per week from flights climbed, blocks walked, and sports played. For each time period, alumni were also categorized by energy expenditure into approximate tertiles: those expending <1,000 kcal per week, those expending 1,000-2,500 kcal per week, and those expending >2,500 kcal per week.

Details on participation in collegiate sports (but not walking or climbing stairs) derived from college archives were available for 4,238 of the 6,092 alumni. These alumni were further categorized by collegiate activity into three levels: nonvarsity athlete playing sports less than 5 hr per week, nonvarsity athlete playing sports 5 hr per week or more, and varsity athlete.

Physical Activities Ranked by MET Score

The MET score is the ratio of work metabolic rate to resting metabolic rate; 1 MET is defined as the energy expended while sitting quietly, which in the average adult is approximately 3.5 ml of oxygen per kilogram of body weight per minute.

MET score = 2.5: boating, sailing; croquet; trail-bike riding

MET score = 3.0: bocci ball, lawn ball; bowling; candlepin bowling, duckpin bowling; carpentry in workshop, do-it-yourself projects; diving; horseshoes; Indian clubs; shuffleboard; surfing, wind surfing

MET score = 3.5: archery; catch, Frisbee, games with children; coaching sports; fishing from riverbank or boat, surf casting; home maintenance, home repair, housekeeping; hunting, shooting, working dogs in field; mowing lawn with power mower

MET score = 4.0: boat maintenance; curling; gardening; golf; raking lawn; walking

MET score = 4.5: badminton; bodybuilding, Nautilius, weight lifting; canoeing for pleasure, kayaking, whitewater rafting; farming, ranching; horseback riding, fox hunting; painting, paper hanging; platform tennis, table tennis; snorkeling, spear fishing; volleyball

MET score = 5.0: baseball, cricket, kickball, softball, track ball, whiffleball; basketball; dancing; digging, spading; gymnastics, trampoline; skateboarding

MET score = 5.5: aerobics, calisthenics, home exercise, tai chi chuan; bicycling; cardiac rehabilitation therapy, health club exercise

MET score = 6.0: bayonet, fencing, kendo, two-handed sword; body surfing, swimming; cross-country hiking; cycling machine, rowing machine, treadmill walking; fishing in stream with wading boots; mowing lawn with hand mower; water skiing

MET score = 6.5: carpentry outside, roof repair, shingling; cutting wood, splitting wood; heavy work around home and yard; paddleball, paddle tennis; sledding, tobogganing; snow shoveling; snow skiing, downhill

MET score = 7.0; backpacking; field hockey, football, lacrosse, rugby; forestry and trail maintenance; ice hockey, ice skating, roller skating; jogging; racquetball; scuba diving; soccer; tennis, court and lawn

MET score = 8.0: boxing; judo, martial arts, wrestling; mountain climbing; polo; snowshoeing; snow skiing, cross-country

From Lee, I.-M. (1992). Time trends in physical activity among college alumni, 1962-1988. *American Journal of Epidemiology*, **108**, 161-175. Baltimore, MD: American Journal of Epidemiology. Reprinted with permission. References 12 and 13 updated since last publication.

MET score = 10.0: handball; logging, lumbering; water polo

MET score = 12.0: canoeing or rowing in competition, racing crew, sculls; competition running, track and field; squash

REFERENCES

12. Ainsworth, B.E., Haskell, W.L., Leon, A.S., Jacobs, D.S., Jr., Montoye, H.J., Sallis, J.F., & Paffenbarger, R.S., Jr. (1993). Compendium of physical activities: Classification of energy costs of human physical activities. *Medicine and Science in Sports and Exercise*, **25**, 71-80.
13. Taylor, H.L., Jacobs, D.R., Jr., Schucker, B., Knudsen, J., Leon, A.S., & DeBacker, G. (1978). A questionnaire for the assessment of leisure time activities. *Journal of Chronic Diseases*, **31**, 741-755.

Physical Activities

1. How many city blocks or their equivalent do you regularly walk each *day*? ___ blocks/day (Let 12 blocks = 1 mile)
2. What is your usual pace of walking? (Please check one.)
   a. ___ Casual or strolling (less than 2 mph) b. ___ Average or normal (2 to 3 mph)
   c. ___ Fairly brisk (3 to 4 mph) d. ___ Brisk or striding (4 mph or faster)
3. How many flights of stairs do you climb *up* each *day*? ____ flights/day (Let 1 flight = 10 steps.)
4. List any sports or recreation you have actively participated in during the past *year*. Please remember seasonal sports or events.

Sport, recreation, or other physical activity	Number of times/year	Average time episode Hr	Min	Years participation
a. ______	______	______	______	______
b. ______	______	______	______	______
c. ______	______	______	______	______
d. ______	______	______	______	______

5. Which of these statements best expresses your view? (Please check one.)
   a. ____ I take enough exercise to keep healthy b. ____ I ought to take more exercise. c. ____ Don't know.
6. At least once a *week,* do you engage in regular activity akin to brisk walking, jogging, bicycling, swimming, etc. long enough to work up a sweat, get your heart thumping, or get out of breath?
   ____ No Why not? ______________ ____ Yes How many times per week? ____ Activity: ______________
7. When you are exercising in your usual fashion, how would you rate your level of exertion (degree of effort)? (Please circle one number.)

0	0.5	1	2	3	4	5	6	7	8	9	10	•
Nothing at all	Very,very weak, (just noticeable)	Very weak	Weak	Moderate	Somewhat strong	Strong (heavy)		Very strong			Very, very strong (almost maximal)	Maximal

8. On a usual weekday and weekend day, how much time do you spend on the following activities? Total for each day should add to 24 hr.

	Usual weekday hr/day	Usual weekend day hr/day
a. Vigorous activity (digging in the garden, strenuous sports, jogging, aerobic dancing, sustained swimming, brisk walking, heavy carpentry, bicycling on hills, etc.)		
b. Moderate activity (housework, light sports, regular walking, golf, yard work, lawn mowing, painting, repairing, light carpentry, ballroom dancing, bicycling on level ground, etc.)		
c. Light activity (office work, driving a car, strolling, personal care, standing with little motion, etc.)		
d. Sitting activity (eating, reading, desk work, watching TV, listening to radio, etc.)		
e. Sleeping or reclining		

From Paffenbarger, R.S., Jr., Blair, S.N., Lee, I.-M., & Hyde, R.T. (1993). Measurement of physical activity to assess health effects in free-living populations. *Medicine and Science in Sports and Exercise*, **25**, 60-70. Reprinted with permission.

Physical Activity Recall Items

Now we would like to know about your physical activity during the past 7 days. But first, let me ask you about your sleep habits.

1. On the average, how many hours did you sleep each night during the last 5 weekday nights (Sunday-Thursday)? ___ hours

2. On the average, how many hours did you sleep each night last Friday and Saturday nights? ___ hours

Now I am going to ask you about your physical activity during the past 7 days, that is, the last 5 weekdays and last weekend, Saturday and Sunday. We are not going to talk about light activities, such as slow walking or light housework, or unstrenuous sports such as bowling, archery, or softball. Please look at this list, which shows some examples of what we consider moderate, hard, and very hard activities. [Interviewer: hand subject card No. 9 and allow time for the subject to read it over.] People engage in many other types of activities, and if you are not sure where one of your activities fits, please ask me about it.

3. First, let's consider moderate activities. What activities did you do and how many total hours did you spend during the last 5 weekdays doing these moderate activities or others like them? Please tell me to the nearest half hour. ___ hours

4. Last Saturday and Sunday, how many hours did you spend on moderate activities and what did you do? (Probe: Can you think of any other sports, job, or household activities that would fit into this category?) ___ hours

5. Now, let's look at hard activities. What activities did you do and how many total hours did you spend during the last 5 weekdays doing these hard activities or others like them? Please tell me to the nearest half hour. ___ hours

6. Last Saturday and Sunday, how many hours did you spend on hard activities and what did you do? (Probe: Can you think of any other sports, job, or household activities that would fit into this category?) ___ hours

7. Now, let's look at very hard activities. What activities did you do and how many total hours did you spend during the last 5 weekdays doing these very hard activities or others like them? Please tell me to the nearest half hour. ___ hours

8. Last Saturday and Sunday, how many hours did you spend on very hard activities and what did you do? (Probe: Can you think of any other sports, job, or household activities that would fit into this category?) ___ hours

9. Compared with your physical activity over the past 3 months, was last week's physical activity more, less, or about the same?

___ 1. More
___ 2. Less
___ 3. About the same

Interviewer: Please list below any activities reported by the subject which you don't know how to classify. Flag this record for review and completion.

Activity (brief description)	Hr: workday	Hr: weekend day
________	________	________
________	________	________
________	________	________

Card No. 9 Examples of Activities in Each Category

Moderate Activity

Occupational tasks: delivering mail or patrolling on foot; house painting; truck driving (making deliveries, lifting and carrying light objects)

From Sallis, J.F., Haskell, W.L., Wood, P.D., Fortmann, S.P., Rogers, T., Blair, S.N., and Paffenbarger, R.S., Jr. (1985). Physical activity assessment methodology in the five-city project. *American Journal of Epidemiology*, **121**, pp. 91-106. Copyright 1985 by The American Journal of Epidemiology. Adapted with permission.

Household tasks: raking the lawn; sweeping and mopping; mowing the lawn with a power mower; cleaning windows

Sports activities (actual playing time): volleyball; Ping-Pong; brisk walking for pleasure or to work (4.83 km/hr [3 miles/hr] or 20 min/km [mile]); golf, walking, and pulling or carrying clubs; calisthenic exercises

Hard Activity

Occupational tasks: heavy carpentry; construction work, doing physical labor

Household tasks: scrubbing floors

Sports activities (actual playing time): tennis doubles; disco, square, or folk dancing

Very Hard Activity

Occupational tasks: very hard physical labor, digging or chopping with heavy tools; carrying heavy loads such as bricks or lumber

Sports activities (actual playing time): jogging or swimming; singles tennis; racquetball; soccer

Note. Some examples of various levels of activity are given. However, the list in Appendix C is more comprehensive and can be used with moderate activities rated 3 to 5 METs; hard activities, 5.1 to 6.9 METs; and very hard activities, 7 METs and more.

Appendix G

Baecke Questionnaire

Questionnaire, Codes, and Method of Calculation of Scores on Habitual Physical Activity

1. What is your main occupation?
   .... 1—3—5
2. At work I sit
   never/seldom/sometimes/often/always .... 1—2—3—4—5
3. At work I stand
   never/seldom/sometimes/often/always .... 1—2—3—4—5
4. At work I walk
   never/seldom/sometimes/often/always .... 1—2—3—4—5
5. At work I lift heavy loads
   never/seldom/sometimes/often/very often .... 1—2—3—4—5
6. After working I am tired
   very often/often/sometimes/seldom/never .... 5—4—3—2—1
7. At work I sweat
   very often/often/sometimes/seldom/never .... 5—4—3—2—1
8. In comparison with others of my own age I think my work is physically
   much heavier/heavier/as heavy/lighter/much lighter .... 5—4—3—2—1
9. Do you play sport?
   yes/no
   If yes:
   —which sport do you play most frequently? .... Intensity 0.76—1.26—1.76
   —how many hours a week? .... <1/1-2/2-3/3-4/>4 Time 0.5—1.5—2.5—3.5—4.5
   —how many months a year? .... <1/1-3/4-6/7-9/>9 Proportion 0.04—0.17—0.42—0.67—0.92

   If you play a second sport:
   —which sport is it? .... Intensity 0.76—1.26—1.76
   —how many hours a week? .... <1/1-2/2-3/3-4/>4 Time 0.5—1.5—2.5—3.5—4.5
   —how many months a year? .... <1/1-3/4-6/7-9/>9 Proportion 0.04—0.17—0.42—0.67—0.92
10. In comparison with others of my own age I think my physical activity during leisure time is
    much more/more/the same/less/much less .... 5—4—3—2—1
11. During leisure time I sweat
    very often/often/sometimes/seldom/never .... 5—4—3—2—1
12. During leisure time I play sport
    never/seldom/sometimes/often/very often .... 1—2—3—4—5
13. During leisure time I watch television
    never/seldom/sometimes/often/very often .... 1—2—3—4—5
14. During leisure time I walk
    never/seldom/sometimes/often/very often .... 1—2—3—4—5
15. During leisure time I cycle
    never/seldom/sometimes/often/very often .... 1—2—3—4—5
16. How many minutes do you walk and/or cycle per day to and from work, school, and shopping?
    <5/5-15/15-30/30-45/>45 .... 1—2—3—4—5

From Baecke, J.A.H., Burema, J., & Fritters, E.R. (1982). A short questionnaire for the measurement of habitual physical activity in epidemiological studies. *American Journal of Clinical Nutrition*, **36**, pp. 936-942.

Calculation of the simple sport-score (I_9):
(a score of zero is given to people who do not play a sport)

$I_9 = \Sigma$ (intensity $\times$ time $\times$ proportion)

= 0/0.01-<4/4-<8/8-<12/≥12 .... 1—2—3—4—5

Calculation of scores of the indices of physical activity:
Work index = $[I_1 + (6 - I_2) + I_3 + I_4 + I_5 + I_6 + I_7 + I_8]/8$
Sport index = $[I_9 + I_{10} + I_{11} + I_{12}]/4$
Leisure time index = $[(6 - I_{13}) + I_{14} + I_{15} + I_{16}]/4$

Appendix H

Health Insurance Plan of New York Questionnaire

Physical Activity Connected With Job—Construction of Index

Question	Answer	Assigned weight
Time on job spent sitting	Practically all	0
	More than half	1
	About half	2
	Less than half	3
	Almost none	4
Time on job spent walking	Almost none	0
	Less than half	1
	About half	2
	More than half	3
	Practically all	4
Walking getting to and from job	None or less than 1 block	0
	1 or 2 blocks	1
	3 or 4 blocks	2
	5 to 9 blocks	3
	10 to 19 blocks	4
	20 to 39 blocks (1 mile, not 2)	5
	40+ blocks (2 + miles)	6
Transportation to and from job	None	0
	Car, bus, railroad, and/or ferry	1
	Subway	2
	Subway + one or more other modes of transportation	3
Lifting or carrying heavy things	Very infrequently or never	0
	Sometimes	3
	Frequently	6
Hours on job	Less than 25	1
	25-34	2
	35-40	3
	41-50	4
	51+	5

Definition of job physical activity classes for males:

Class	Accumulated weight
I	1-10
II	11-14
III	15-18
IV	19-28

Note: Individuals failing to answer one or more components of the job activity questionnaire are not classified.

Reprinted from the *Journal of Chronic Diseases*, **18**. The H.I.P. study of incidence and prognosis of coronary heart disease, Shapiro, S., Weinblatt, E., Frank, C.W., & Sager, R.V., pages 527-558, with kind permission from Pergamon Press Ltd, Headington Hill Hall, Oxford OX3 0BW, UK.

Physical Activity Off the Job—Construction of Index

Question	Frequently	Sometimes	Very infrequently or never
Take walks in good weather	2	1	0
Work around house or apartment	2	1	0
Gardening in spring or summer	2	1	0
Take part in sports			
Active ball game other than golf, bowling, pool or billiards	4	3	0
Other	3	2	0

Definition of off-job physical activity classes:

Class	Accumulated weight
1	0-1
2	2-3
3	4-5
4	6-10

Note: Individuals failing to answer one or more components of the off-job activity questionnaire are not classified.

British Civil Servant Questionnaire

To the extent memory permits, we want you to give us a complete record of all you did on Friday and Saturday. Anything of 5 min duration or more which you remember doing during these 2 days should be listed. An example of part of a completed record is attached; please study it. We call your attention to the following points:

- The record should show what actually happened on Friday and Saturday. Please do not describe a typical day.
- The whole day should be accounted for, from getting up in the morning till retiring for the night, and there should be no time-gap.
- Every activity should be listed in the order in which it occurred during the day. The hours are provided for you to record the items according to time of day.

The description of each activity or pastime need only be brief. Please avoid general terms like ''pottering about,'' ''housecleaning,'' ''gardening''; give specific descriptions such as ''repairing an electrical plug,'' ''washing windows,'' ''weeding in garden.'' Time spent at your office need only be described as ''work.'' However, indicate how you spent your lunch time or any other breaks during working hours.

We would like details of any journey you took during the day, including your journey to and from work, indicating the amount of time spent walking, in transport and type of transport involved. For example:

7:05 p.m.	Journey to visit friends	
	Walk to station	5 min
	Stand waiting at station	5 min
	Sit in train	15 min
	Walk to friend's house	5 min

If you report an activity as lasting more than half an hour, try to remember whether you took a break or were interrupted during this time. If so, you should show this on the record thus:

2:05 p.m.	Digging in garden (5 min break for chat and smoke)	2 hr, 20 min

In situations where you were moving back and forth among several tasks and cannot easily indicate at what time you were doing each, please list the main task involved and indicate the approximate amount of time spent at the subtasks. For example:

5:10 p.m.	Working in garden	1 hr
	(digging, 30 min)	
	(lawn mowing, 30 min)	

When you have finished the records for Friday and Saturday, please answer the additional questions which follow.

Please try to complete your questionnaire today, during working hours if at all possible.

Then follow similar sheets for Friday afternoon (noon-6:00 p.m.), Friday evening (6:00 p.m.-midnight), Saturday morning (6:00 a.m.-noon), Saturday afternoon (noon-6:00 p.m.), and Saturday evening (6:00 p.m.-midnight). There is also an additional sheet for recording activities carried out after midnight on Friday and Saturday.

Confidential Activity Record for Friday ______________________

Morning: 6:00 a.m. to noon

Suggestions: a. Start with beginning of the day and continue through it chronologically.
b. Use pencil; have eraser handy for amending entries where necessary.
c. Please write legibly.

(1) Time	(2) Time activity began to nearest 5 min	(3) Please enter here brief descriptions of what you were doing in line with "Time activity began" in Column 2	Duration of activity to nearest 5 min
6:00 a.m.			
to			
7:00 a.m.			
7:00 a.m.			
to			
8:00 a.m.			
8:00 a.m.			
to			
9:00 a.m.			
9:00 a.m.			
to			
10:00 a.m.			
10:00 a.m.			
to			
11:00 a.m.			
11:00 a.m.			
to			
noon			

↑ Hourly time slots may be ignored providing the correct time is shown in Column 2 opposite each entry.

Confidential

For each question below please tick one answer most applicable to you

1. Compared to the typical Friday at this time of year, was the past Friday

   1. ❑ *Less physically active than usual* 2. ❑ *About average for physical activity* 3. ❑ *More physically active than usual*

2. Compared to the typical Saturday at this time of year, was the past Saturday

   1. ❑ *Less physically active than usual* 2. ❑ *About average for physical activity* 3. ❑ *More physically active than usual*

3. On most Saturdays or Sundays the average person in your age group spends one hour at each of the following activities: walking, gardening, household chores, and "do-it-yourself" projects. Compared to such a person how physically active do you consider yourself?

   1. ❑ *Very active* 2. ❑ *Fairly active* 3. ❑ *Average* 4. ❑ *Fairly inactive* 4. ❑ *Very inactive*

   (Tick one category only, please)

4. Please tick the category which best answers each of the following questions.

   (i) In general, I am usually tense or nervous.
   This describes me
   1. ❑ *Very well* 2. ❑ *Fairly well* 3. ❑ *Not very well* 4. ❑ *Not at all*

   (ii) There is a great amount of nervous strain connected to my daily activities.
   This describes me
   1. ❑ *Very well* 2. ❑ *Fairly well* 3. ❑ *Not very well* 4. ❑ *Not at all*

   (iii) At the end of the day I am completely exhausted, mentally and physically.
   This describes me
   1. ❑ *Very well* 2. ❑ *Fairly well* 3. ❑ *Not very well* 4. ❑ *Not at all*

   (iv) My daily activities are extremely trying and stressful.
   This describes me
   1. ❑ *Very well* 2. ❑ *Fairly well* 3. ❑ *Not very well* 4. ❑ *Not at all*

5. All things considered, how do you feel about your present job? (Tick one)

   1. ❑ *I like it very much*
   2. ❑ *I like it a good deal*
   3. ❑ *I like it*
   3. ❑ *I like it a little*
   5. ❑ *I am indifferent to it*
   6. ❑ *On the whole I don't like it*
   7. ❑ *I dislike it*
   8. ❑ *I dislike it very much*

5a. At work about how much of your time do you usually spend sitting down?

   1. ❑ *All or nearly all* 2. ❑ *More than half* 3. ❑ *Less than half* 1. ❑ *None, almost none*

Framingham Leisure Time Physical Activity Questionnaire

REFERENCES

Folsom, A.R., Caspersen, C.J., Taylor, H.L., Jacobs, D.R., Jr., Luepker, R.V., Gomez-Marin, O., Gillum, R.F., & Blackburn, H. (1985). Leisure time physical activity and its relationship to coronary risk factors in a population-based sample. The Minnesota Heart Survey. *American Journal of Epidemiology*, **121**, 570-579.

Fox, S.M., Naughton, J.P., & Gorman, P.A. (1972). Physical activity and cardiovascular health. *Modern Concepts of Cardiovascular Disease*, **41**, 17-30.

Passmore, R. & Durnin, J.V.G.A. (1955). Human energy expenditure. *Physiological Reviews*, **35**, 801-840.

Taylor, H.L., Jacobs, D.R., Schucker, B., Knudsen, J., Leon, A.S., & DeBacker, G. (1978). A questionnaire for the assessment of leisure time physical activities. *Journal of Chronic Diseases*, **31**, 741-755.

From Dannenberg, A.L. & Wilson, P.W.E. (1989). Framingham leisure time physical activity questionnaire. In T.M. Drury (Ed.) *Assessing Physical Fitness and Physical Activity in Population-Based Surveys*, pp. 650-651. DHHS Publication No. (PHS) 89-1253, Washington, DC: U.S. Government Printing Office.

BUMC -Framingham Study Exam 2 Code sheet	Numerical lab data Deck 221	Date this exam ______ Date last exam ______

Cols.	Code	Item
1-4		Record number / Name / Age (years)
5-6		How many flights of stairs do you climb up/day? (10 steps = 1 flight)
7-8		How many city blocks do you walk/day? (1 mi = 12 blocks)
	A. Code / Hr / Min	Sports or recreation you have participated in during past week
9-14		1.
15-20		2.
21-26		3.
	A. Code / No weeks/ year / Hr / Min	Sports or recreation you have participated in during past year
27-34		1.
35-42		2.
43-50		3.
51-58		4.
59-66		5.
67-74		6.
75-76		How many times a week do you engage in any regular sports or recreational activity with sufficient intensity to work up a sweat?
77	Less active 0 / More active 1 / Same 2 / Unknown 9	Would you say that during the past week you were less active, more active, or about the same as normal?
78	0 / 1 / 2 / 9	How would you relate your physical activities compared to others?
79		In your occupation how many hours a day are spent doing: Vigorous activity (working up a sweat)
80		Moderate activity (increased breathing, not sweating)
81		Light activity (sitting, driving, slowly walking)
120-122	2 2 1	Deck number

Kilocalories Assignments for Framingham Leisure Time Physical Activities in Order of Activity Codes

Source	kcal · min⁻¹	Code	Activity	Source	kcal · min⁻¹	Code	Activity
M	3.5	1	Walking for pleasure	M	12	42	Soccer
M	4	2	Walking to work	M	3.5	43	Golf—power cart
M	8	3	Using stairs	M	5	44	Golf—pulling cart
M	6	4	Cross-country hiking	M	5.5	45	Golf—carrying clubs
M	7	5	Backpacking	M	2.5	46	Mowing lawn—ride mower
M	8	6	Mountain climbing/hill climbing	M	4.5	47	Mowing lawn—power mower
M	4	7	Bicycling—work, pleasure	M	6	48	Mowing lawn—push mower
M	6	8	Dancing	M	4.5	49	Weeding, gardening
M	4.5	9	Home exercise	M	5	50	Spading, digging
M	6	10	Health club exericse	M	4	51	Raking lawn
M	6	11	Jog/walk	M	6	52	Snow shoveling
M	12	12	Running < 10 min/mile	F	7.5	53	Wood chopping
M	6	13	Weight lifting	M	3	54	Carpentry—inside
M	6	14	Water skiing	M	4.5	55	Painting—inside
M	3	15	Sailing—competition	M	6	56	Carpentry—outside
M	3.5	16	Canoeing—pleasure	M	5	57	Painting—oustide
M	12	17	Canoeing—competition	M	3.5	58	Fishing—bank or boat
F	6	18	Kayaking, rafting	M	6	59	Fishing—wading
M	6	19	Swimming in a pool	M	6	60	Hunting—pheasants, grouse
M	6	20	Swimming at a beach	M	5	61	Hunting—small game
M	7	21	Scuba diving	M	6	62	Hunting—large game
M	5	22	Snorkeling	F	9	64	Hockey
M	7	23	Skiing—downhill	M	3.5	66	Horseback riding
M	8	24	Skiing—cross-country	B	5	67	Karate, martial arts
M	7	25	Skating	B	5	68	Car washing
M	7	26	Sledding	M	8	70	Snowshoeing
M	3	27	Bowling	B	4	71	Whiffle ball
M	4	28	Volleyball	B	5	73	Frisbee
M	4	29	Table tennis	F	2.5	74	Dirt-bike riding
M	6	30	Tennis—singles	F	3.5	75	Car and boat repair
M	6	31	Tennis—doubles	B	8	77	Wood carrying
M	5	32	Softball, baseball	M	3	78	Sailing—pleasure
M	7	33	Badminton	B	5	80	Coaching sports
M	8	34	Paddleball	B	5	82	Playing with children
M	10	35	Racquetball	B	8	83	Lacrosse
M	6	36	Basketball—nongame	P	5	85	Golf ball hitting
M	8	37	Basketball—game	B	5	90	Yard and barn chores
M	7	38	Basketball—officiating	B	5	93	Lobstering, clamming
M	8	39	Touch football	B	5	94	House building
M	10	40	Handball	B	5	95	Umpiring, referee
M	10	41	Squash	B	6	98	Wind surfing, board sailing

Sources: M—Minnesota: Taylor et al. and Folsom et al.; F—Fox et al.; P—Passmore and Durnin; B—Best estimate.

U.S. National Health Survey, 1991, Physical Activity Questionnaire

These next questions are about physical exercise.	**Section L — Physical Activity and Fitness**	3-4
ITEM L1 *Mark from observation or previous information.*	1☐ SP is physically handicapped *(Describe in footnotes, then 1)* 2☐ Other *(2)*	5
1a. In the past 2 weeks (outlined on that calendar), beginning Monday *(date)* and ending this past Sunday *(date)*, have you done any exercises, sports, or physically active hobbies?	1☐ Yes (1b) 2☐ No 3☐ Don't know (DK) } *(3 on page 40)*	6
b. What were they? *Record on next page, then 1c.*		
c. Anything else?	☐ Yes *(Reask 1b and c)* ☐ No*(2b)*	

Notes

Section L — Physical Activity and Fitness

Note: *Ask all of 2a before going to 2b—d.*			***Note:*** *Ask 2b — d for each activity marked "Yes" in 2a.*		
Hand calendar. 2a. In the past 2 weeks (outlined on that calendar), beginning Monday *(date)* and ending this past Sunday *(date)*, have you done any of the following exercises, sports, or physically active hobbies —	Yes / No		b. How many times in the past 2 weeks did you [go/do] *(activity in 2a)*?	c. On the average, about how many minutes did you actually spend (doing) *(activity in 2a)* each time?	d. {What usually happened to your heart rate or breathing when you [did/went] *(activity in 2a)*} Did you have a small, moderate, or large increase, or no increase at all in your heart rate or breathing?
(1) Walking for exercise?	1☐ 2☐	7	(1)____ times 8-9	____ min 10-12	1☐Small 3☐Large 2☐Moderate 0☐None 9☐DK 13
(2) Gardening or yard work?	1☐ 2☐	14	(2)____ times 15-16	____ min 17-19	1☐Small 3☐Large 2☐Moderate 0☐None 9☐DK 20
(3) Stretching exercises?	1☐ 2☐	21	(3)____ times 22-23	____ min *(next activity)* 24-26	
(4) Weightlifting or other exercises to increase muscle strength?	1☐ 2☐	27	(4)____ times 28-29	____ min 30-32	1☐Small 3☐Large 2☐Moderate 0☐None 9☐ DK 33
ITEM L2 *Refer to age*	☐ *75 + {2a (19)}* ☐ Other *{2a(5)}*				
(5) Jogging or running?	Yes / No 1☐ 2☐	34	(5)____ times 35-36	____ min 37-39	1☐Small 3☐Large 2☐Moderate 0☐None 9☐DK 40
(6) Aerobics or aerobic dancing?	1☐ 2☐	41	(6)____ times 42-43	____ min 44-46	1☐Small 3☐Large 2☐Moderate 0☐None 9☐DK 47
(7) Riding a bicycle or exercise bike?	1☐ 2☐	48	(7)____ times 49-50	____ min 51-53	1☐Small 3☐Large 2☐Moderate 0☐None 9☐ DK 54
(8) Swimming laps or water exercises?	1☐ 2☐	55	(8)____ times 56-57	____ min 58-60	1☐Small 3☐Large 2☐Moderate 0☐None 9☐ DK 61
(9) Playing tennis?	1☐ 2☐	62	(9)____ times 63-64	____ min 65-67	1☐Small 3☐Large 2☐Moderate 0☐None 9☐ DK 68
(10) Bowling	1☐ 2☐	69	(10)____ times 70-71 *(next activity)*		
(11) Playing golf	1☐ 2☐	72	(11)____ times 73-74 *(next activity)*		

Section L — Physical Activity and Fitness - *(Continued)*

Hand calendar. In the past 2 weeks (outlined on that calendar), beginning Monday *(date)* and ending this past Sunday *(date)*, have you done any of the following exercises, sports, or physically active hobbies —		b. How many times in the past 2 weeks did you [go/do] *(activity)*?	c. On the average, about how many minutes did you actually spend (doing) *(activity)* each time?	d. {What usually happened to your heart rate or breathing when you [did/went] *(activity)*} Did you have a small, moderate, or large increase, or no increase at all in your heart rate or breathing?
ITEM L3 *Refer to age*	☐ 65 + *{2a (19)}* ☐ Other *{2a(12)}*			
(12) Playing baseball or softball?	Yes 1☐ No 2☐ 75	(12)____ times 76-77	____ min 78-80	1☐Small 2☐Moderate 3☐Large 0☐None 9☐DK 81
(13) Playing handball, racquetball, or squash?	1☐ 2☐ 82	(13)____ times 83-84	____ min 85-87	1☐Small 2☐Moderate 3☐Large 0☐None 9☐DK 88
ITEM L4 *Refer to age*	☐ 50 + *{2a (19)}* ☐ Other *{2a(14)}*			
(14) Skiing?	☐ Yes ☐ No *(15)* 88	90-91		
(a) Downhill?	1☐ 2☐	(a)____ times *(next activity)*		
(b) Cross-country?	1☐ 2☐ 92	(b)____ times 93-94	____ min 95-97	1☐Small 2☐Moderate 3☐Large 0☐None 9☐DK 98
(c) Water?	1☐ 2☐ 99	(c)____ times 100-101 *(next activity)*		
(15) Playing basketball?	1☐ 2☐ 102	(15)____ times 103-104	____ min 105-107	1☐Small 2☐Moderate 3☐Large 0☐None 9☐DK 108
(16) Playing volleyball ?	1☐ 2☐ 109	(16)____times 110-111	____ min 112-114	1☐Small 2☐Moderate 3☐Large 0☐None 9☐DK 115
ITEM L5 *Refer to age*	☐ 40 + *{2a (19)}* ☐ Other *{2a(17)}* RT 32 3—4			
(17) Playing soccer ?	1☐ 2☐ 5	(17)____ times 6-7	____ min 8-10	1☐Small 2☐Moderate 3☐Large 0☐None 9☐DK 11

Section L — Physical Activity and Fitness - *(Continued)*

Hand calendar. In the past 2 weeks (outlined on that calendar), beginning Monday *(date)* and ending this past Sunday *(date)*, have you done any of the following exercises, sports, or physically active hobbies —	b. How many times in the past 2 weeks did you [go/do] *(activity)*?	c. On the average, about how many minutes did you actually spend doing *(activity)* each time?	d. {What usually happened to your heart rate or breathing when you [did/went] *(activity)* Did you have a small, moderate, or large increase, or no increase at all in your heart rate or breathing?
ITEM L6 *Refer to age* ☐ 30 + *{2a (19)}* ☐ Other *{2a(18)}*			
(18) Playing football 1☐ 2☐ [12]	(18)____ times [13-14]	____ min [15-17]	1☐Small 2☐Moderate 3☐Large 0☐None 9☐DK [18]
(19) Have you done any (other) exercises, sports, or physically active hobbies in the past 2 weeks? ☐ Yes - What were they? ☐ No Anything else? *If listed activity, mark "yes" for that activity, otherwise, specify below.* ____________ [19-20]	(19)____ times [21-22]	____ min [23-25]	1☐Small 2☐Moderate 3☐Large 0☐None 9☐DK [26]
____________ [27-28]	(19)____ times [29-30]	____ min [31-33]	1☐Small 2☐Moderate 3☐Large 0☐None 9☐DK [34]

Appendix L

Modified Baecke Questionnaire

Questionnaire, Codes, and Method of Calculation of Scores on Habitual Physical Activity in Elderly People

Household activities

1. Do you do the light household work (dusting, washing dishes, repairing clothes, etc.)?
   0. Never (<once a month)
   1. Sometimes (only when partner or help is not available)
   2. Mostly (sometimes assisted by partner or help)
   3. Always (alone or together with partner)
2. Do you do the heavy housework (washing floors and windows, carrying trash disposal bags, etc.)?
   0. Never (<once a month)
   1. Sometimes (only when partner or help is not available)
   2. Mostly (sometimes assisted by partner or help)
   3. Always (alone or together with partner)
3. For how many persons do you keep house (including yourself; fill in "0" if you answered "never" in Q1 and Q2)?
4. How many rooms do you keep clean, including kitchen, bedroom, garage, cellar, bathroom, ceiling, etc.? (Fill in "0" if you answered "never" in Q1 and Q2.)
   1. 1-6 rooms
   2. 7-9 rooms
   3. 10 or more rooms
5. If any rooms, on how many floors? (Fill in "0" if you answered "never" in Q4.)
6. Do you prepare warm meals yourself, or do you assist in preparing?
   0. Never
   1. Sometimes (once or twice a week)
   2. Mostly (3-5 times a week)
   3. Always (more than 5 times a week)
7. How many flights of stairs do you walk up per day? (One flight of stairs is 10 steps.)
   0. I never walk stairs
   1. 1-5
   2. 6-10
   3. More than 10
8. If you go somewhere in your hometown, what kind of transportation do you use?
   0. I never go out
   1. Car
   2. Public transportation
   3. Bicycle
   4. Walking

From Voorrips, L.E., Ravelli, A.C.J., Dongelmans, P.C.A., Deurenberg, P., & VanStaveren, W.A. (1991). A physical activity questionnaire for the elderly. *Medicine and Science in Sports and Exercise*, **23**, pp. 974-979. Copyright 1991 by Wiliams & Wilkins. Reprinted with permission.

Questionnaire, Codes, and Method of Calculation of Scores on Habitual Physical Activity in Elderly People *(Continued)*

9. How often do you go out for shopping?
   0. Never or less than once a week
   1. Once a week
   2. Twice to four times a week
   3. Every day
10. If you go out for shopping, what kind of transportation do you use?
   0. I never go out for shopping
   1. Car
   2. Public transportation
   3. Bicycle
   4. Walking

Household score = (Q1 + Q2 + . . . + Q10)/10

Sport activities
Do you play a sport?

Sport 1: Name

Intensity (code)	(1a)
Hours per week (code)	(1b)
Period of the year (code)	(1c)

Sport 2: Name

Intensity (code)	(2a)
Hours per week (code)	(2b)
Period of the year (code)	(2c)

Sport score $\sum_{i=1}^{2} (ia \bullet ib \bullet ic)$

Leisure time activities
Do you have other physically active activities?

Activity 1: Name

Intensity (code)	(1a)
Hours per week (code)	(1b)
Period of the year (code)	(1c)

Activities 2 to 6: as activity 1.

Leisure time activity score $\sum_{i=1}^{6} (ja \bullet jb \bullet jc)$

Questionnaire score = household score + sport score = Leisure time activity score

Codes:
Intensity code[1]:

0: Lying, unloaded	Code 0.028
1: Sitting, unloaded	Code 0.146
2: Sitting, hand or arm movements	Code 0.297
3: Sitting, body movements	Code 0.703
4: Standing, unloaded	Code 0.174
5: Standing, hand or arm movements	Code 0.307
6: Standing, body movements, walking	Code 0.890
7: Walking, hand or arm movements	Code 1.368
8: Walking, body movements, cycling, swimming	Code 1.890

Hours per week:	
1: Less than 1 hr • wk^{-1}	Code 0.5
2: [1,2 > hr • wk^{-1}	Code 1.5
3: [2,3 > hr • wk^{-1}	Code 2.5
4: [3,4 > hr • wk^{-1}	Code 3.5
5: [4,5 > hr • wk^{-1}	Code 4.5
6: [5,6 > hr • wk^{-1}	Code 5.5
7: [6,7 > hr • wk^{-1}	Code 6.5
8: [7,8 > hr • wk^{-1}	Code 7.5
9: More than 8 hr • wk^{-1}	Code 8.5

Months per year:	
1: Less than 1 month • year^{-1}	Code 0.04
2: 1-3 months	Code 0.17
3: 4-6 months	Code 0.42
4: 7-9 months	Code 0.67
5: More than 9 months • year^{-1}	Code 0.92

[1]Unitless intensity code, originally based on energy costs.

Appendix M

Zutphen Physical Activity Questionnaire

The questions asked in this part concern your daily activities.

1. Can you walk indoors?
   - ❏ Yes
   - ❏ No, because I use a wheelchair.
   - ❏ No, because I am bedridden.
   - ❏ No, for a different reason, namely ________

   If you are unable to walk/cycle, you can go on to question number 7.

2. How often did you take a walk during the last week?

   _______ times

3. How long did such a walk last?

   _______ minutes

4. How would you describe your walking pace?
   - ❏ Calm
   - ❏ Normal
   - ❏ Firm

5. Did you take a walk that lasted longer than 1 hour during the last month?
   - ❏ No
   - ❏ Yes

5a. How often did you do that?

   _______ times

6. Do you bicycle?
   - ❏ No
   - ❏ Yes

6a. If yes, how often did you bicycle last week?

   _______ times

6b. How long did you bicycle?

   _______ minutes

6c. How would you describe your bicycling pace?
   - ❏ Calm
   - ❏ Normal
   - ❏ Fast

7. Do you have a garden?
   - ❏ No
   - ❏ Yes

7a. If yes, how many hours, on average, a week do you spend in your garden?

   In summer _______ hours

   In winter _______ hours

7b. Do you work in your garden by yourself?
   - ❏ No
   - ❏ Yes
   - ❏ Partly

8. Do you do the odd jobs in and around the house by yourself (e.g., painting and carpentry)?
   - ❏ No
   - ❏ Yes

8a. If yes, for how many hours a month?

   _______ hours

9. Did you participate in sports lately?
   - ❏ No
   - ❏ Yes

From Casperson, C.J., Bloemberg, B.P.M., Saris, W.H.M., Merritt, R.K., & Kromhout, D. (1991). The prevalence of selected physical activities and their relation with coronary heart disease risk factors in elderly men: The Zutphen study, 1985. *American Journal of Epidemiology*, **133**, pp. 1078-1092.

9a. If yes, what kind of sport?

____

9b. How many hours, on average, do you spend participating in sports monthly?

❑ Less than 1 hour a month

❑ _______ hours a month

10. Do you have a hobby (other than gardening or sports)?

❑ No
❑ Yes

10a. If yes, what kind of hobby?

____

10b. How many hours a week do you spend on it?

❑ Less than 1 hour a week
❑ _______ hours a week

11. Did you keep birds as pets for a least 1 year during the past 10 years?

❑ No
❑ Yes

11a. If yes, what kind of birds?

❑ Canaries
❑ Parrots
❑ Pigeons
❑ Chickens
❑ Other

12. How often did you perspire during physical exercise in the last week?

❑ Never
❑ _______ times

13. Do you climb stairs regularly?

❑ No
❑ Yes

14. Most men of your age spend about 1 hour a day doing domestic work, doing odd jobs, gardening, walking, and doing other physical activities. How do you see yourself compared with these men?

❑ Far more active
❑ More active
❑ About the same
❑ Less active
❑ Far less active

15. What do you think of your pace compared with men of your age?

❑ Much faster
❑ Faster
❑ About the same
❑ Slower
❑ A lot slower

16. How many hours, on average, do you sleep at night?

_______ hours

17. How many hours do you sleep during the day (for example, take a nap)?

❑ I do not sleep in the daytime.

❑ I sleep for _______ hours.

Intensity Codes for Hobbies and Sports

Activity	Intensity code	Activity	Intensity code
Babysitting	1.5	Photography or film	1.5
Badminton	5.5	Playing shuffleboard	2.0
Billiards	2.3	Politics	1.5
Bicycling	4.5	Puzzles	1.5
Bookkeeping	1.5	Reading	1.2
Bowling	3.5	Running	6.0
Cinema	2.0	Sailing	3.0
Club activities	1.8	Sauna	1.8
Collecting	1.5	Sculpture	3.2
Cooking	2.0	Skating	7.0
Cross-country skiing	7.5	Skiing	7.2
Dancing	5.0	Soccer	7.0
Drawing	1.5	Socializing	1.8
Driving	1.4	Sports	3.0
Fancy needlework	1.3	Studying	1.4
Fishing	2.3	Swimming	5.0
Former occupation	—[a]	Table tennis	4.0
Games	1.5	Taking care of plants	1.8
Gardening	4.5	Tennis	6.0
Golf	3.5	Training dogs	2.2
Gymnastics	4.0	Traveling	1.8
Hockey	7.5	Television or radio	1.3
Horseback riding	3.5	Volunteer work	2.0
Hunting	3.5	Walking	3.5
Keeping animals or birds	1.8	Water sports	3.0
Music or acting	2.4	Yoga	1.8
Odd jobs	3.5		

[a]The intensity code for former occupation varied with the type of work activity performed.

Appendix N

The Yale Physical Activity Survey

Interviewer, please mark time: __ __ : __ __ : __ __
Hr Min Sec

> Interviewer: (Please hand the subject the list of activities while reading this statement.) Here is a list of common types of physical activities. Please tell me which of them you did during a *typical week in the last month.* Our interest is learning about the types of physical activities that are a part of your *regular work and leisure routines.*
>
> For each activity you do, please tell me how much time (in hours) you spent doing this activity during a typical week. (Hand subject card #1.)

	Time (hr/week)	Intensity code (kcal/min)
Work		
Shopping (e.g., grocery, clothes)	______	3.5
Stair-climbing while carrying a load	______	8.5
Laundry (time loading, unloading, hanging, folding only)	______	3.0
Light housework: tidying, dusting, sweeping; collecting trash in home; polishing; indoor gardening; ironing	______	3.0
Heavy housework: vacuuming, mopping; scrubbing floors and walls; moving furniture, boxes, or garbage cans	______	4.5
Food preparation (10+ min in duration): chopping, stirring; moving about to get food items, pans	______	2.5
Food service (10+ min in duration): setting table; carrying food; serving food	______	2.5
Dishwashing (10+ min in duration): clearing table; washing/drying dishes; putting dishes away	______	2.5
Light home repair: small appliance repair; light home maintenance/repair	______	3.0
Heavy home repair: painting, carpentry, washing/polishing car	______	5.5
Other: ____________________	______	______
Yard work		
Gardening: planting, weeding, digging, hoeing	______	4.5
Lawn mowing (walking only)	______	4.5
Clearing walks/driveway: sweeping, shoveling, raking	______	5.0
Other: ____________________	______	______

Caretaking

Older or disabled person (lifting, pushing wheelchair)	_______	5.5
Childcare (lifting, carrying, pushing stroller)	_______	4.0

Exercise

Brisk walking (10+ min in duration)	_______	6.0
Pool exercises, stretching, yoga	_______	3.0
Vigorous calisthenics, aerobics	_______	6.0
Cycling, exercycle	_______	6.0
Swimming (laps only)	_______	6.0
Other: ______________________________	_______	_______

Recreational activities

Leisurely walking (10+ min in duration)	_______	3.5
Needlework: knitting, sewing, needlepoint	_______	1.5
Dancing (moderate/fast): line, ballroom, tap, square	_______	5.5
Bowling, bocci	_______	3.0
Golf (walking to each hole only)	_______	5.0
Racquet sports: tennis, racquetball	_______	7.0
Billiards	_______	2.5
Other: ______________________________	_______	_______

> Interviewer: (Please read statement to subject.) I would now like to ask you about certain types of activities that you have done during *the past month.* I will ask you about how much vigorous activity, leisurely walking, sitting, standing, and other things that you usually do.

1. About how many times during the month did you participate in *vigorous* activities that lasted at least *10 minutes* and caused large increases in breathing, heart rate, or leg fatigue *or* caused you to perspire? (Hand subject card #2.)

   Score: 0 = Not at all (go to Question 3)
   1 = 1-3 times per month
   2 = 1-2 times per week
   3 = 3-4 times per week
   4 = 5+ times per week
   7 = Refused
   8 = Don't know

   Frequency score = ____________

2. About how long do you do this vigorous activity each time? (Hand subject card #3.)

   Score: 0 = Not applicable
   1 = 10-30 min
   2 = 31-60 min
   3 = 60+ min
   7 = Refused
   8 = Don't know

   Duration score = ____________

   Weight = 5

Vigorous activity index score:

Frequency score ____ × duration score ____ × weight ____ = __________
(Responses of 7 or 8 are scored as missing.)

3. Think about the walks you have taken during the past month. About how many times per month did you walk for *at least 10 minutes* or more *without stopping* which *was not* strenuous enough to cause large increases in breathing, heart rate, or leg fatigue *or* cause you to perspire? (Hand subject card #2.)

   0 = Not at all (go to Question 5)
   1 = 1-3 times per month
   2 = 1-2 times per week
   3 = 3-4 times per week
   4 = 5+ times per week
   7 = Refused
   8 = Don't know

   Frequency score = ______________

4. When you did this walking, for how many minutes did you do it? (Hand subject card #3.)

   Score: 0 = Not applicable
   1 = 10-30 min
   2 = 31-60 min
   3 = 60+ min
   7 = Refused
   8 = Don't know

   Duration score = ______________

   Weight = 4

Leisurely walking index score:

Frequency score ____ × duration score ____ × weight ____ = __________
(Responses of 7 or 8 are scored as missing.)

5. About how many hours a day do you spend moving around on your feet while doing things? Please report only the time that you are *actually moving*. (Hand subject card #4.)

   Score: 0 = Not at all
   1 = Less than 1 hr per day
   2 = 1 to less than 3 hr per day
   3 = 3 to less than 5 hr per day
   4 = 5 to less than 7 hr per day
   5 = 7+ hr per day
   7 = Refused
   8 = Don't know

   Moving score = ______________

   Weight = 3

Moving index score:

Moving score ____ × weight ____ = __________
(Responses of 7 or 8 are scored as missing.)

6. Think about how much time you spend standing or moving around on your feet on an average day during the past month. About how many hours per day do you *stand*? (Hand subject card #4.)

   Score: 0 = Not at all
   1 = Less than 1 hr per day
   2 = 1 to less than 3 hr per day
   3 = 3 to less than 5 hr per day
   4 = 5 to less than 7 hr per day
   5 = 7+ hr per day
   7 = Refused

8 = DK

Standing score = ____________

Weight = 2

Standing index score:

Standing score _____ × weight _____ = ____________
(Responses of 7 or 8 are scored as missing.)

7. About how many hours did you spend sitting on an average day during the past month? (Hand subject card #5.)

Score: 0 = Not at all
1 = Less than 3 hr
2 = 3 hr to less than 6 hr
3 = 6 hr to less than 8 hr
4 = 8+ hr
7 = Refused
8 = DK

Sitting score = ____________

Weight = 1

Sitting index score:

Sitting score _____ × weight _____ = ____________
(Responses of 7 or 8 are scored as missing.)

8. About how many flights of stairs do you climb *up* each day? (Let 10 steps = 1 flight.) ____________

9. Please compare the amount of physical activity that you do during other seasons of the year with the amount of activity you just reported for a typical week in the past month. For example, in the summer, do you do more or less activity than what you reported doing in the past month?

(Interviewer, please circle the appropriate score for each season.)

	Lot more	Little more	Same	Little less	Lot less	Don't know
Spring	1.30	1.15	1.00	0.85	0.70	—
Summer	1.30	1.15	1.00	0.85	0.70	—
Fall	1.30	1.15	1.00	0.85	0.70	—
Winter	1.30	1.15	1.00	0.85	0.70	—

Seasonal adjustment score = sum over all seasons / 4 ____________

Interviewer, please mark time: __ __ : __ __ : __ __
Hr Min Sec

Card #1

WEEKLY PHYSICAL ACTIVITIES

Work

Shopping (e.g., grocery, clothes)	
Stair climbing while carrying a load	
Laundry	
Light Housework:	tidying, dusting, sweeping, collecting garbage in home, polishing, indoor gardening, ironing
Heavy Housework:	vacuuming, mopping, scrubbing floors and walls, moving furniture, moving boxes or garbage cans
Food preparation (10+ min):	chopping, stirring, moving around to get food items and pots or pans
Food service (10+ min):	setting table, carrying food, serving food
Dishwashing (10+ min):	clearing table, washing and drying dishes, putting dishes away
Light home repair:	small appliance repair, light household maintenance and repair tasks
Heavy home repair:	painting, washing and polishing car, carpentry
Other:	______________________________

Yardwork

Gardening:	pruning, planting, weeding, hoeing, digging
Lawn mowing (walking only)	
Clearing walks and driveway:	raking, shoveling, sweeping
Other:	______________________________

Caretaking

Older or disabled person:	lifting, pushing wheelchair
Child care:	lifting, pushing stroller

Exercise

Brisk walking for exercise (10+ min):	causes large increases in heart rate, breathing or leg fatigue
Stretching exercises, yoga, pool exercise	
Vigorous calisthenics, aerobics:	causes large increases in heart rate, breathing or leg fatigue
Cycling, exercycle	
Lap swimming	
Other:	______________________________

Recreational Activities

Leisurely walking (10+ min)	
Hiking	
Needlework:	knitting, sewing, crocheting, needlepoint
Dancing (mod/fast):	line dancing, ballroom, square, tap, etc.
Bowling, boccie	
Golf (walking to each hole only)	
Racquet sports:	tennis, racquetball
Billiards	
Other:	______________________

Card #2

Not at all
1-3 times per month
1-2 times per week
3-4 times per week
5 or more times per week
Don't know

Card #3

10-30 minutes
31-60 minutes
60 or more minutes
Don't know

Card #4

Not at all
less than 1 hour per day
1 to less than 3 hours per day
3 to less than 5 hours per day
5 to less than 7 hours per day
7 or more hours per day
Don't know

Card #5

Not at all
less than 3 hours per day
3 hours to less than 6 hours per day
6 hours to less than 8 hours per day
8 or more hours per day
Don't know

Appendix O

Netherlands Health Education Project Questionnaire

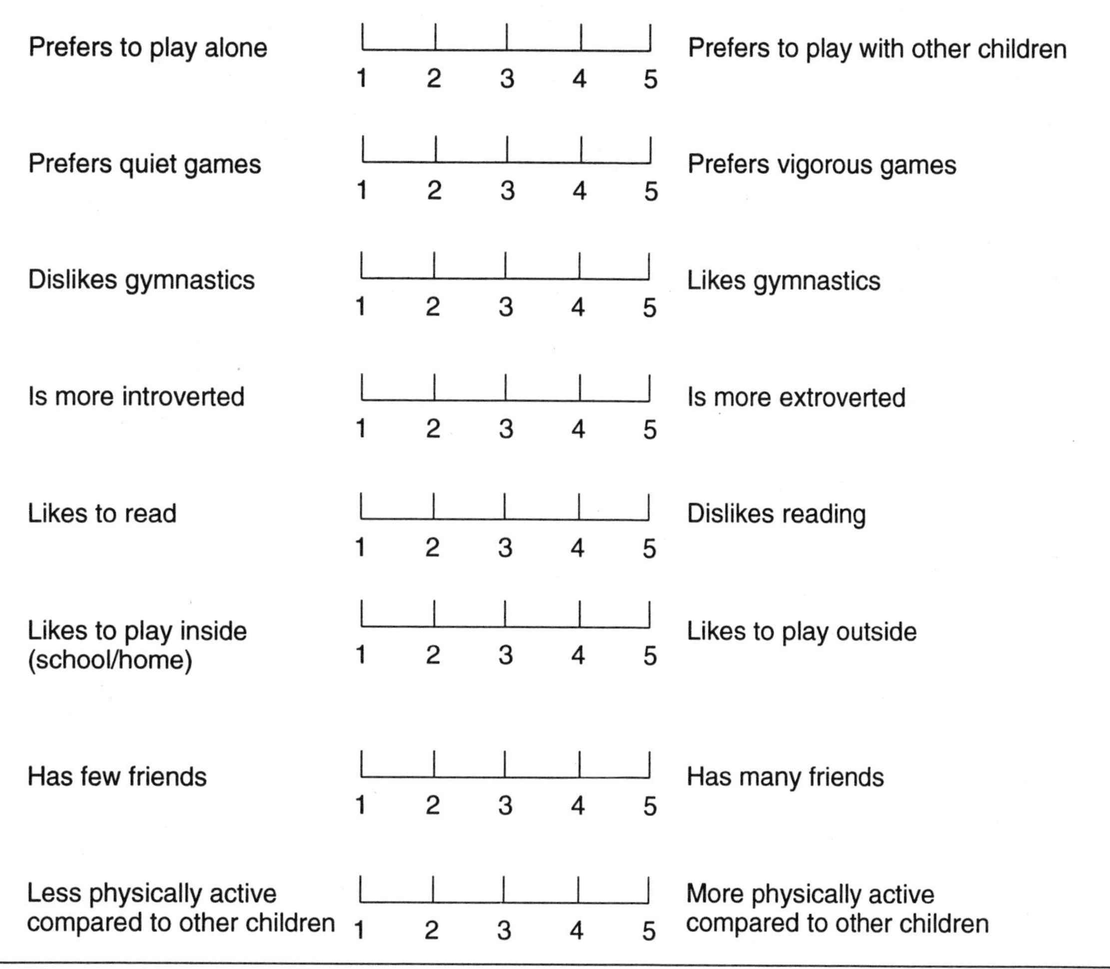

Prefers to play alone	1	2	3	4	5	Prefers to play with other children
Prefers quiet games	1	2	3	4	5	Prefers vigorous games
Dislikes gymnastics	1	2	3	4	5	Likes gymnastics
Is more introverted	1	2	3	4	5	Is more extroverted
Likes to read	1	2	3	4	5	Dislikes reading
Likes to play inside (school/home)	1	2	3	4	5	Likes to play outside
Has few friends	1	2	3	4	5	Has many friends
Less physically active compared to other children	1	2	3	4	5	More physically active compared to other children

Each item is scored from 1 to 5. The Physical Activity Score is then computed as follows:

$$\text{PA Score} = \frac{\Sigma \text{ 8 items}}{40} \bullet 100\%$$

From Saris, W.C.M., Doesburg, W.H., Lemmens, W.A.J.G., & Reingis, A. (1974). *Habitual physical activity in children: Results of a questionnaire and movement counters*, (pp. III79-III92). Niymegen, The Netherlands: Report of the Health Education Project (GVO).

Appendix P

The Amsterdam Growth Study Questionnaire

Main Items of the Activity Questionnaire/Interview

Activities in Relation to School or Work

I. Activities in relation to school or work
  A. All transportation
    Means: 1st and 2nd possible choice (walking, cycling, or public transport) and average number of times used
    Active time per means (both ways)
    Score = mean active transportation time in minutes per week (intensity = 1)
  B. Physical education and extracurricular sport activity
    Average active time per lesson
    Obligatory number of lessons per week
    Real number of lessons participated in
    Real extracurricular activity
    Score = mean active physical education or sport time in minutes per week (intensity = 2)
  C. Work-related activities like walking, cycling, lifting, and stair climbing
    Average active time per working day above minimum intensity
    Score = mean activity plus active transportation time in minutes per week

II. Organized activities
  A. Activities in sport clubs
    Kind of sport club membership
    1. Training (club 1; repeat in case of more memberships)
      Average active time per training
      Number of training sessions per week
      Real number of training sessions participated in
      Score = mean active training time in minutes per week (intensity = 1, 2, or 3)
    2. Matches (club 1; repeat in case of more memberships)
      Average active playing time per match
      Number of matches per week
      Real number of matches participated in
      Score = mean active playing time in minutes per week (intensity = 1, 2, or 3)
    3. Transportation
      Score = mean active transportation time in minutes per week of all organized sports activities (intensity = 1)
  B. Activities in other clubs
    Kind of membership
    Average active time per meeting above minimum intensity (club 1)
    Real number of meetings participated in
    Score = mean activity plus active transportation time in minutes per week (intensity = 1)

From Kemper, H.C.G. (1995). *The Amsterdam growth study: A longitudinal analysis of health, fitness, and lifestyle*, (pp. 42-44). Champaign, IL: Human Kinetics. Reprinted with permission.

III. Unorganized activities

A. General leisure time activities (unorganized sports, gardening, etc.)
Monday through Sunday: average active time in minutes per week (intensity = 1, 2, or 3; see Table)
Score = mean active leisure time in minutes per week (intensity = 1, 2, or 3; see Table)

B. Activities in jobs (housekeeping, shopping, voluntary work)
Kind of job
Average active time per working day above minimum intensity
Score = mean activity plus active transportation time in minutes per week (intensity = 1)

Classification of the Intensity Level of Work, Sports, and Leisure Activities Into Four Categories on the Basis of Their Average Intensity

Intensity		Activity
Very light < 4 METs	Domestic:	Washing dishes, dusting, sweeping floors
	Outdoor:	Sitting, standing, strolling
	Sport:	Billiards, bowling, bridge, checkers, chess, cricket, fishing, gliding, golf, sailing, shooting, skittle, t'ai chi ch'uan
Light 4-7 METs (Level 1)	Domestic:	Beating carpets, carrying groceries, hammering, polishing floors, sawing, scrubbing floors, lifting
	Outdoor:	Bicycling, canoeing, rowing, walking
	Sport:	Ballet, baseball, bodybuilding, dancing (ballroom, modern, folk), gymnastics (rhythmic, remedial, jazz), hiking, horseback riding, softball, table tennis, tug-of-war, volleyball, waterskiing, weight lifting
Medium-heavy 7-10 METs (Level 2)	Domestic:	Stair climbing
	Outdoor:	Basketball (dribbling, shooting), playing active games, skating, soccer (dribbling, kicking), swimming
	Sport	Track and field (field events), badminton, fencing, gymnastics, mountaineering, scuba diving, skating (figure, speed), skiing (alpine), tennis (outdoor, indoor)
Heavy ≥ 10 METs (Level 3)	Outdoor:	Basketball (game), running, soccer (game)
	Sport:	Track and field (track events), basketball, canoeing, conditioning exercises, cycling (race), handball (European; indoor, outdoor), hockey (field, ice, roller; indoor, outdoor), jogging, kick-boxing, netball (indoor, outdoor), martial arts (judo, jujitsu, karate, aikido, kendo, kung fu, tae kwan do), rowing, rugby, skiing (cross-country), soccer (indoor, outdoor), squash, swimming, trampolining, water polo, wrestling

Note. Domestic = at home; Outdoor = unorganized recreational activity; Sport = activity in sport clubs.

INDEX

A
Abbreviations and symbols, 119-120
Accelerometers, portable, 79-89. *See also* Caltrac accelerometers; single-plane accelerometers
  actometers, 57, 76, 77, 80-81, 82
  diary method compared to, 82, 87
  doubly labeled water method compared to, 86, 89
  force plates/platforms compared to, 79, 83
  heart rate method compared to, 84, 85, 86, 107-108, 109, 110
  photographic methods compared to, 79
  questionnaires/interviews compared to, 46, 49, 51, 52-53, 55, 56, 57, 82, 87-88
  respiration chambers compared to, 85, 89
  summary evaluation of, 118
  triaxial, 85-86, 88-89, 109, 110
Actigraph, 86
Activity, muscular, estimating energy expenditure from, 10-12, 79, 83
Activity monitoring. *See* Large-Scale Integrated Motor Activity Monitor
Actometers, 57, 76, 77, 80-81, 82
Aerobic power (VO_2max), in validation
  of assessment methods in general, 5
  of movement assessment devices, 81, 86
  of physiologic response methods, 107-108, 110
  of questionnaires/interviews, 43, 47, 48, 49, 50, 51, 52, 54, 55, 56, 61
Aerobics Institute, 53
Ainsworth, B.E., 45, 53, 60
Alameda County (Calif.) physical activity study, 53, 54
Albanes, D., 52
Alcohol consumption, 19, 20
Alderson, M.R., 50
AMF Quantum XL, 100
Amsterdam growth study questionnaire, 57, 58, 62, 109-110, 182-183
Anastasiades, P., 109
Andersen, K.L., 34, 50
Andrews, R.B., 103
Anorexia nervosa, 20
Antonelli, D., 79
Atwater, W.O., 7
Australia, questionnaire/interview-based studies in, 54
Avons, P., 81
Ayen, T.G., 86, 88

B
Baecke, J.A.H., 50, 99, 103-104. *See also* Baecke questionnaire
Baecke questionnaire, 50, 54-55, 62, 154-155, 169-171
  compared to other questionnaires, 46, 47, 49, 51, 52
  with elderly, 54-55, 62, 169-171
  movement assessment devices compared to, 78, 87
Baker, J.A., 99
Baranowski, T., 32, 56
Barber, C., 75
Bar-Or, O., 102
Bassey, E.J., 73, 102, 109
Behavioral observation methods, 26-33, 118
  compared to other methods, 60, 77, 78, 82, 86, 107-108
Belgium, questionnaire/interview-based studies in, 53
Bell, R.Q., 80
Benedict, F.G., 7, 73, 98
Berkowitz, R.I., 77
Bhattacharya, A., 79, 84
Bicycle racing, 20
Binkhorst, R.A., 26, 81, 99, 103-104
Biomechanical methods, 10-12, 79, 83
Biomotometer, 76
Black, Joseph, 6
Black, R.A., 43
Blair, D., 34
Blair, S.N., 45
Blood pressure, 97
Blood pressure method, 55, 56
Bodwell, C.E., 19
Body size, and energy expenditure, 4
Body temperature method, 97
Boltwood, Bertram Borden, 17
Bone area/density, 53, 54
Borg Scale, 43
Bouchard, C., 34
Breast-feeding, 21
British Civil Servant questionnaire, 50, 62, 158-160
British Thermal Units (BTUs), 4, 120
Brody, S., 7
Brooks, C.M., 53
Brouha, L., 11
Brown, A.E., 99
Brown, G.W., 11
Brück, K., 105
Bryant, J.C., 102
BTUs (British Thermal Units), 4, 120
Bullen, B.A., 27
Buono, M.J., 60, 101-102
Burema, J., 50
Burn patients, 21
Buskirk, E.R., 34

C
Caloric intake. *See* Energy consumption (dietary intake)
Calories, 4, 120
Calorimeters, 6-10, 11, 12, 39, 105. *See also* Respiration chambers
Calorimetry, direct, 6-7. *See also* Indirect (respiratory) calorimetry; oxygen consumption (VO_2), measurement of
  compared to other methods, 39, 105, 107-108
Caltrac accelerometers, 81-83, 84, 85, 86, 87-88
  behavioral observation compared to, 82, 86
  compared to other movement assessment devices, 78, 81, 82, 86, 88
  diary method compared to, 82, 87
  doubly labeled water method compared to, 86
  heart rate method compared to, 86, 107-108

questionnaires/interviews compared to, 46, 49, 51, 52-53, 55, 56, 82, 87-88
Campbell, Joan, 48
Canada, questionnaire/interview-based studies in, 53, 57
Canadian Fitness Survey, 53
Canary, J.J., 18
Carbohydrate utilization, 7-8
Carbon dioxide production, measurement of, 7, 8, 9, 18, 122
Cardiovascular disease, correlation with questionnaires/interviews, 57. *See also* Coronary heart disease
Carlson, D., 101-102
Caspersen, C.J., 44, 45, 116
Centers for Disease Control Behavioral Risk Surveillance, 53
Cerebral palsy, 21
Children
behavioral observation method with, 26, 27, 28-30, 31-32, 60
diary method with, 34, 35, 38, 40
doubly labeled water (DLW) method with, 21
energy costs of activities of, 123
heart rate method with, 60, 100, 101-102, 105
movement assessment devices with, 60, 76, 77, 80, 86
questionnaires/interviews with, 56-59, 60, 62, 181-183
Cinematography. *See* Photography; videotaping
Closed circuit method, 8
Colburn, T.R., 76, 83
Cole, T.J., 18, 19
Computers
for analyzing laboratory photography, 10-11
with behavioral observation method, 27, 30-31, 32
with heart rate method, 109
with motion counters, 76-77
with questionnaires/interviews, 44
Connett, J., 45
Conservation of energy, principle of, 6
Continuous heart rate recorders, 99-102
Conversion table, 120
Conway, J.M., 18
Coronary heart disease, and risk factors
Large-Scale Integrated Motor Activity Monitor correlated with, 77
questionnaires/interviews correlated with, 44, 45, 50, 51, 52, 54, 55, 57-58
Cotes, J.E., 73, 79
Coward, W.A., 18, 19
Cramer, D.B., 99
Crawford, P.B., 26
Cunningham, D.M., 11

D

Dagerman, R., 80
Dallosso, H.M., 73
Datamyte 1002, 32
Dauncy, M.J., 105
Da Vinci, Leonardo, 72
Dawson, W.W., 81
De Looze, M.P., 12
Denmark, questionnaire/interview-based studies in, 53
Dewhurst, D.J., 83
Diabetes, 53, 58
Diary method, 34-41
with children, 34, 35, 38, 40
compared to other methods, 34, 35, 39, 46, 49, 78, 82, 87, 106
with elderly, 35
with questionnaires/interviews, 56
summary evaluation of, 118
Dictionary of Occupational Titles, 44
Dietary balance method. *See* Energy consumption (dietary intake), measurement of
Dietary induced thermogenesis, 4
Digital recording system, for behavioral observation, 30
Distance conversion table, 120
Doesburg, W.H., 57
Doubly labeled water (DLW) method, 17-25
accelerometers compared to, 86, 89
advantages and disadvantages of, 17, 21
assumption and sources of error with, 18-19
with children, 21
diary method compared to, 39
energy consumption measurement compared to, 12, 19, 20
heart rate method compared to, 21, 105, 106, 107-108
history of, 17-18
indirect (respiratory) calorimetry compared to, 18, 19, 20
procedures for, 17, 18
questionnaires/interviews compared to, 46, 48, 49, 51, 55
recent applications of, 20-21
respiration chambers compared to, 18, 19, 20
summary evaluation of, 117, 118
validation of, 19-20, 118
for validation of other methods in general, 5
Douglas bag method, 8-10
Dowley, E.M., 80
Durant, R.H., 57-58,, 102
Duration, definition of, 3
Durnin, J.V.G.A., 34

E

Eberhart, H.J., 79
ECGs (electrocardiograms), 98-100, 109
Edholm, O.G., 27, 34
Elderly adults
Baecke questionnaire with, 54-55, 62, 169-171
diary method with, 35
energy costs of activities of, 123
Physical Activity Scale for the Elderly (PASE) with, 55-56
Yale Physical Activity Survey (YPAS) with, 55, 56, 62, 175-180
Zutphen questionnaire with, 55, 62, 172-174
Electrocardiocorders, 34
Electrocardiograms (ECGs), 98-100, 109
Electromyography, 98, 109
Electronic calorie counter, 83
Electronic motion sensors. *See* Movement assessment devices
Elftman, H., 11
Elgin National Watch Company, 99
Ellis, M.J., 26
Elvers, J.W.H., 104
EMGs (electromyograms), 98, 109
Emons, H.J.G., 105
Endurance athletes/training, 21
Energy consumption (dietary intake), measurement of, 12
diary method compared to, 39
doubly labeled water (DLW) method compared to, 12, 19, 20
heart rate method compared to, 106, 107-108
questionnaires/interviews compared to, 46, 48, 49, 50, 51, 52
Energy costs of activities. *See also* METs
table of, 123-132
use with diary method, 34-35
Energy expenditure
definition of, 4
need for measures of, 3, 116-117
units for expression of, 4, 120
Energy/heat conversion table, 120
Evans, A., 109
Evans, D., 75
Event counters, for behavioral observation, 30
Exersentry heart rate recorder, 100

F

Face validity, definition of, 57. *See also* Coronary heart disease, and risk factors
Factorial method, 35
Fales, E., 26, 32

Fargo Activity Timesampling Survey (FATS), 31-32
Fat utilization, 7-8
Fentem, P.H., 73, 75, 102
Fick equation, 98
Field assessment methods. *See also* Behavioral observation methods; diary method; doubly labeled water (DLW) method; movement assessment devices; physiologic response to activity methods; questionnaires and interviews
  Douglas bag method as, 9
  for energy consumption measurement, 12
  summary of evaluation of, 117, 118
Finland, questionnaire/interview-based studies in, 53, 57
First law of thermodynamics, 6
Fitness measures, in validation. *See also* Aerobic power (VO_2max), in validation
  of assessment methods in general, 5
  of questionnaire/interview method, 43, 47, 48, 49, 51, 52, 53, 54, 55, 56, 61
Five-city/7-day recall questionnaire, 45, 48-50, 62, 152-153
  with children, 57, 58, 59, 62
  compared to other questionnaires, 46, 47, 49, 51, 52
  movement assessment devices compared to, 49, 87
Flow-through technique, 8
Folsom, A.R., 45
Food consumption. *See* Energy consumption (dietary intake)
Food sources, and oxygen consumption, 7-8
Force plates/platforms, 11-12, 79, 83
Forslund, A., 7
Foster, F.G., 30
Framingham, Massachusetts questionnaire, 50-51, 62, 161-163
  compared to other questionnaires, 46, 47, 49, 51, 52
Franz, H., 9
Freedson, P.S., 60
Frequency, definition of, 3
Friedman, M., 99
Frisch, R.E., 53
Frÿtters, J.E.R., 50

G

Gallup Poll, 53
Gambian study subjects, 20
Giauque, William Frances, 17
Gilliam, T.B., 26
Gima, A.S., 43
Glagov, S., 99, 102
Goldfield, S.R.W., 45
Goldsmith, R., 104
Gomez-Marin, O., 45
Goran, M.I., 18
Gordon, G.B., 17
Göteborg, Sweden questionnaire, 52
Gray, R.G., 57-58
Great Britain, questionnaire/interview-based studies in, 53
Greenleaf, J.E., 79
Greer, J.R., 98
Gregory, J., 50
Gretebeck, R., 81
Gretebeck, R.J., 102
Grimby, G., 54, 57
Groenenboom, D.C., 105
Guarini, J.J., 76, 83
Guatemalan study subjects, 20

H

Hagen, J.W., 80
Haldane apparatus, 8
Hale, T., 104
Halverson, C.F., Jr., 80, 86
Hambraeus, L., 7
Hanish, H.M., 101
Harkness, J.W., 57-58
Hartman, T.J., 45
Harvard University physical activity study, 53
Haskell, W.L., 109
Hawes, M.R., 58
Heady, J.A., 43
Health and Welfare Canada survey, 53
Health Insurance Plan of New York questionnaire (HIP), 50, 62, 156-157
  Caltrac accelerometer compared to, 87
  compared to other questionnaires, 46, 47, 49, 51, 52
Heartbeat accumulators, 99
Heart disease. *See* Coronary heart disease, and risk factors
Heart rate distribution recorders, 99
Heart rate method, 98-105, 106-108
  with adults, 102-105
  behavioral observation compared to, 107-108
  with behavioral observation method, 31
  calorimetry compared to, 105, 107-108
  with children, 60, 100, 101-102, 105
  diary method compared to, 106
  doubly labeled water method compared to, 21, 105, 106, 107-108
  energy consumption measurement compared to, 106, 107-108
  instrumentation, 98-101
  movement assessment devices compared to, 76, 77, 84, 85, 86, 107-108, 109, 110
  net heart rate in, 104
  oxygen consumption compared to, 98, 99, 101, 102-105, 109
  questionnaires/interviews compared to, 60, 107-108
  reproducibility of, 101-102, 118
  summary evaluation of, 118
Heat/energy conversion table, 120
Heat production, measurement of. *See* Calorimetry, direct
Hebestreit, H., 102
Hedley, O.F., 44
Hegge, F.W., 86
Helmholtz, Hermann Ludwig Ferdinand von, 6
Herron, R.E., 75
HIP. *See* Health Insurance Plan of New York questionnaire
Hiruta, S., 10
Historical (lifetime) physical activity questionnaire, 53, 62
Holter, N.J., 98
Hovell, M.F., 32
HR. *See* Heart rate method
Huenemann, R.L., 26
Humphrey, J.E., 99
Hyde, R.T., 45
Hyperactivity, in children, 80

I

IDECG group, 19
Ikegami, H., 10
Ikegami, Y., 10
Illinois Bell Telephone, 99
IMP (integrating motor pneumotachograph), 9-10
Indirect (respiratory) calorimetry, 7-10, 11, 121-122. *See also* Respiration chambers
  compared to other methods, 18, 19, 20, 85, 89
Indirect validation, definition of, 57. *See also* Coronary heart disease, and risk factors
Inman, N., 79
Integrating motor pneumotachograph (IMP), 9-10
Intensity, definition of, 3
International System of Units (SI), 4
Interviews. *See* Questionnaires and interviews
Irby, P.J., 109
Ireland, questionnaire/interview-based studies in, 53, 57, 62
Irving, J.M., 73, 74
Ismail, A.H., 11
Isotopes
  definition of, 17

in doubly labeled water (DLW) method, 17-19

J

Jacobs, David R., Jr., 44, 45, 50, 52-53
James, W.P.T., 105
Janz, K.F., 85
Jefferson, Thomas, 72
Job, activity on the, 4, 43-44, 118
Job classification method, 43-44, 118
Johnston, D.W., 109
Johnstone, 17
Joule, James Prescott, 6
Joules, 4, 120

K

Keefe, F.J., 80
Kemper, Han C.G., 57, 74, 75, 109-110
Kendrick, J.S., 44
Kilocalories
  conversion of oxygen consumption to, 7
  definition of, 4, 120
  Framingham questionnaire activity values, 163
  in intensity codes with YPAS and Zutphen questionnaires, 55, 174, 175-176
  METs and, 4 (*see also* METs)
  scores generated by Caltrac accelerometer, 81, 82
Kilojoules, 4, 7, 120. *See also* Kilocalories
Klein, P.D., 19
Klesges, R.C., 31-32
Knudson, J., 45
Kofranyi, E., 9
Kofranyi-Michaelis respirometer, 9
Koper, H., 88-89, 109
Kripke, D.F., 83
Kriska, A.M., 53
Kriska lifetime physical activity questionnaire, 53, 62
Ku, L.C., 26
Kupfer, D.J., 30
Kupst, M.J., 76

L

Laboratory methods, 6-14. *See also* Energy consumption (dietary intake), measurement of
  biomechanical methods, 10-12, 79, 83
  direct measurement, 6-10
Lactation, 21
LaGrange, F., 4
Laplace, Pierre-Simon de, 6
LaPorte, R.E., 57, 77, 79
Large-Scale Integrated Motor Activity Monitor (LSI), 77-79
  behavioral observation compared to, 78
  Caltrac accelerometers compared to, 78, 81, 82, 86
  diary method compared to, 78
  questionnaires/interviews compared to, 46, 51, 53, 57, 78
Laughlin, M.E., 43
Lauru, L., 11
Lavoisier, Antoine Laurent, 6
Lee, I.-M., 45
Léger, L., 100
Lemmens, W.A.J.G., 57
Leon, A.S., 45, 53, 60
Leonardo da Vinci, 72
Lewis, Gilbert Newton, 17
Lifetime physical activity questionnaire (Kriska), 53, 62
Lifson, N., 17, 19
Linder, C.W., 57-58
Linear variable differential transformers, 11
Lipid Research Clinics questionnaire, 52-53, 62
  compared to other methods, 52, 53, 87
  compared to other questionnaires, 46, 47, 49, 51, 52
Lipsey, 83
Livingston, M.B.E., 103, 105
London School of Hygiene and Tropical Medicine, 55
LSI. *See* Large-Scale Integrated Motor Activity Monitor
Ludvigsson, J., 58

M

Maccoby, E.E., 80
MacDonald, I.A., 102
Mahoney, O.M., 58
Mansourian, P., 99
Marey, Étienne Jules, 72, 73
Marr, J.W., 50
Marsden, J.P., 75
Marti, B., 57
Masironi, R., 34, 50, 99
Mass spectrometers, 18
Mayer, Julius Robert von, 4, 6, 27
McClintock, R., 17
McCutcheon, E.P., 79
McKenzie, T., 26
McKenzie, T.L., 56
McKinlay, J.B., 45
McKinty, C., 102
McPartland, R.J., 30
Meade, F., 73, 79
Meade, T.W., 50
Medilog 4-24, 109
Meijer, G.A., 88-89
Meijer, G.A.L., 109
Messin, S., 83
Metabolic disorders, 20. *See also* Diabetes
METs
  with accelerometers, 81, 89
  definition of, 4
  with diary method, 35
  with heart rate method, 105
  with questionnaires/interviews, 44, 57, 134, 139, 145, 149-150, 183
  table of, 123-132
  use for assessment summarized, 4
Michaelis, H.F., 9
Micro-Scholander apparatus, 8
Miles, C., 19
Miller Lite Report on American Attitudes Toward Sport, The, 53
Minilogger, 77
Minnesota Leisure Time Activity Questionnaire, 44-45, 62, 134, 139-148
  with children, 57
  compared to other methods, 46, 47, 78, 87
  compared to other questionnaires, 46, 47, 49, 51, 52, 54
Mitra, K.P., 4
Miyamura, M., 10
Monark 1 heart rate recorder, 100
Montgomery, S.R., 75
Montoye, Henry J., 11, 44, 60, 81, 86, 88, 102, 103
Mor, V., 54
Morrell, E.M., 80
Morris, J.N., 43, 50, 55
Morris, J.R.W., 79
Motion counters, 75-79. *See also* Large-Scale Integrated Motor Activity Monitor
Mountain climbing, 20
Movement assessment devices, 72-96. *See also* Accelerometers, portable; Large-Scale Integrated Motor Activity Monitor; pedometers and other step counters
  actometers, 57, 76, 77, 80-81, 82
  behavioral observation compared to, 77, 78, 82, 86
  with children, 60, 76, 77, 80, 86
  diary method compared to, 78, 82, 87
  doubly labeled water method compared to, 86, 89
  force plates/platforms compared to, 79, 83
  heart rate method compared to, 76, 77, 84, 85, 86, 107-108, 109, 110

motion counters, 75-79
oxygen consumption compared to, 76, 78, 79, 81, 84, 85, 88
questionnaires/interviews compared to, 46, 48, 49, 51, 53, 54, 55, 56, 57, 60, 77, 78, 82, 87-88
summary evaluation of, 118
Mt. Everest, climbing of, 20
Mueller, J.K., 102
Mullaney, D.J., 83
Müller, E.A., 9, 98
Müller-Franz calorimeter, 9
Murlin, J.R., 98
Murschhauser, H., 73
Muscular activity, estimating energy expenditure from, 10-12, 79, 83
Myelodysplasia, 21

N
Nader, P., 26
Nader, P.R., 56
Nagy, K.A., 19
Nair, K.S., 18
Nakajima, K., 84-85
National Center for Health Statistics, 3
National Institutes of Health, 3, 76
National Survey of Personal Health Practices in the United States, 53
National surveys, 53, 164-168
Nelson, J.A., 101-102
Net heart rate, 104
Netherlands, questionnaire/interview-based studies in, 53, 54, 57
Amsterdam growth study questionnaire, 57, 58, 62, 109-110, 182-183
Netherlands Health Education Project questionnaire, 57, 62, 181
Zutphen questionnaire, 55, 62, 172-174
Neustein, R.A., 101
New England Research Institute, 55-56
Nicoud, J.N., 99
Nier, A.O., 17, 18
Nitrogen, excreted in urine, 7, 8
Nitrogen, exhaled, 8
Northern Ireland questionnaire, 57, 62
Norway, questionnaire/interview-based studies in, 53
Nutrition, and doubly labeled water (DLW) method, 20, 21

O
Obesity, 20, 21
Observation methods. *See* Behavioral observation methods
Occupational physical activity, 4, 43-44, 118
Open circuit method, 8-10, 121-122
Osana, M., 84-85
Oxygen, discovery of, 6
Oxygen consumption (VO_2), definition of, 7
Oxygen consumption (VO_2), measurement of. *See also* Aerobic power; indirect (respiratory) calorimetry; respiration chambers
body size and, 4
body temperature method compared to, 97
with diary method, 34, 35
diary method compared to, 39
doubly labeled water method compared to, 20
force plate measurements compared to, 11-12
with Framingham questionnaire, 51
heart rate method compared to, 98, 99, 101, 102-105, 109
methods and procedures, 7-10, 121-122
movement assessment devices compared to, 76, 78, 79, 81, 84, 85, 88
ventilation method compared to, 98

P
Pacer 2000H heart rate recorder, 100
Paffenbarger, Ralph S., Jr., 43, 45. *See also* Paffenbarger/Harvard Alumni Questionnaire
Paffenbarger/Harvard Alumni Questionnaire, 45, 62, 149-151
compared to other methods, 48, 78, 87
compared to other questionnaires, 46, 47, 49, 51, 52, 53
Page, R.G., 99
Panasonic pedometer, 72-73
Parks, J.W., 43
PASE (Physical Activity Scale for the Elderly), 55-56
Passmore, R., 34
Patrick, J.M., 73, 74, 102
Pattison, D.C., 50
Pearl, Raymond, 44
Pedometers and other step counters, 72-75, 76, 77, 118
compared to other methods, 51, 54, 57, 76, 77
compared to other movement assessment devices, 76, 79
Perrier Survey, The, 53
Perry, J., 79
Photography, for behavioral observation in the field, 27
Photography, for laboratory estimation of energy expenditure, 10-11, 12
compared to other methods, 79, 107-108
Physical activity, definition of, 3
Physical activity levels, dimensions and means of expressing, 3-4
Physical Activity of Finnish Adults survey, 53
Physical Activity Scale for the Elderly (PASE), 55-56
Physical Activity Score (Netherlands Health Education Project), 57, 181
Physical environment, effect on response to activity, 3
Physiologic response to activity methods, 97-115. *See also* Heart rate method
individual methods, 97-98
multiple methods, 109-110
multiple recording systems, 105, 109
Piezoelectric crystals/devices, 11, 76, 81, 83, 88
Portable accelerometers. *See* Accelerometers, portable
Portable calorimeters, 9, 10, 11
Portable respirometers, 8-10, 11
Post-Gorden, J.G., 80
Powell, K.E., 44
Pregnancy, 21
Prentice, A.M., 19
President's Council on Physical Fitness and Sports survey, 53
Priestly, Joseph, 6
Protein utilization, 7-8
Psychological/emotional effects on response to activity, 3-4
Puerto Rico, questionnaire/interview-based studies in, 53
Punch card observation system, 27

Q
Questionnaires and interviews, 42-71
advantages and limitations of, 42-43
basic guidelines, 42
behavioral observation compared to, 60
with children, 56-59, 60, 62, 181-183
compared to other questionnaires, 46, 47, 49, 51, 52, 53, 54
diary method compared to, 46, 49
doubly labeled water method compared to, 46, 48, 49, 51, 55
with elderly adults, 54-56, 62, 169-180
energy consumption measurement compared to, 46, 48, 49, 50, 51, 52
fitness measures compared to, 43, 47, 48, 49, 51, 52, 53, 54, 55, 56, 61
heart rate method compared to, 60, 107-108
for job classification, 43-44
movement assessment devices compared to, 46, 48, 49, 51, 52-53, 54, 55, 56, 57, 77, 78, 82, 87-88
summary evaluation of, 118
with young/middle-aged adults, 43-54, 62
specific methods and instruments. *See also* Baecke questionnaire; Five-city/7-day recall questionnaire; Minnesota Leisure Time Activity Questionnaire; Paffenbarger/Harvard Alumni Questionnaire
Amsterdam growth study, 57, 58, 62, 109-110, 182-183
British Civil Servant, 50, 62, 158-160
Framingham, Massachusetts, 46, 47, 49, 50-51, 52, 62, 161-163
Göteborg, Sweden, 52
Health Insurance Plan of New York, 46, 47, 49, 50, 51, 52, 62, 87, 156-157
Kriska lifetime physical activity, 53, 62
Lipid Research Clinics, 46, 47, 49, 51, 52-53, 62, 87

national health surveys, 53, 164-168
Netherlands Health Education Project, 57, 62, 181
Northern Ireland, 57, 62
Physical Activity Scale for the Elderly (PASE), 55-56
Tecumseh, 44-45, 46, 47, 52, 62, 87, 134-139, 145
Yale Physical Activity Survey (YPAS), 55, 56, 62, 88, 175-180
Zutphen, 55, 62, 172-174

R

Racey, P.A., 19
Raffle, P.A.B., 43
Ramsden, R.W., 75
Rauramaa, R., 45
Redmond, D.P., 86
Reed, R.B., 27
Reeh, J.J., 98
Regnault, Henri-Victor, 8
Reiff, G.G., 4, 44
Reingis, A., 57
Reiset, 8
Reisman, J.M., 80
Report of the Working Group on Heart Disease Epidemiology, 3
RER (respiratory exchange ratio), 7
Respiration chambers, 7-8
compared to other methods, 18, 19, 20, 85, 89
Respiratory calorimetry. *See* Indirect (respiratory) calorimetry
Respiratory exchange ratio (RER), 7
Respiratory quotient (RQ), 7, 8, 9, 19
Respirometers, portable, 8-10, 11
Resting metabolic rate, definition of, 4
Reswick, J.B., 79
Richardson, J.F., 102
Riddell, M., 102
Riddoch, C., 57
Ritmeester, J.W., 57
Riumallo, J.A., 34
Roberts, C.G., 43
Roby, J.J., 101-102
Room calorimeters, 7, 12, 39, 105
Rosenman, R.H., 99
Rowley, D.A., 99
Roy, P., 4
RQ (respiratory quotient), 7, 8, 9, 19
Running, doubly labeled water (DLW) method applied to, 21
Rutenfranz, J., 34, 50

S

Sallis, J., 26
Sallis, J.F., 48, 58, 60, 101-102
Saltin, B., 54, 57
Saltin and Grimby questionnaire, 54, 57
SAMI (Socially Accepted Monitoring Instrument), 99
Sanders, R.T., 101
Saris, Wim H.M., 26, 57, 81, 99, 103-104, 105
Saunders, J., 79
Schoeller, D.A., 18, 19, 20
Schulman, J.L., 76, 80
Schultz, S., 105
Seale, J.L., 18, 19
Seiko 1 heart rate recorder, 100
Seliger, F., 34
Seliger, V., 50
Serial event timer and recorder (SETAR), 27
Servais, S.B., 11
SETAR (serial event timer and recorder), 27
7-day recall questionnaire. *See* Five-city/7-day recall questionnaire
Shapiro, L.R., 26
Shvarta, E., 79
SI (International System of Units), 4
Sickness Impact Profile, 55
Simmons, N.W., 76, 83
Single-plane accelerometers, 79-85, 87-88
diary method compared to, 82, 87
doubly labeled water method compared to, 86
force plates/platforms compared to, 79, 83
heart rate method compared to, 84, 85, 86, 107-108
questionnaires/interviews compared to, 46, 49, 51, 52-53, 82, 87-88
Sjödin, A., 7
Smith, B.M., 76, 83
Smith, K.W., 45
Snel, P., 99
Socially Accepted Monitoring Instrument (SAMI), 99
Soddy, Frederick, 17
SOFIT system, 26
Soldiers in training, 21
Speakman, J.R., 18, 19
Spectrometers, 18
Spina bifida, 20
Sports Tester PE 3000, 100
Spurr, G.B., 105
Stanford Five-City Project. *See* Five-city/7-day recall questionnaire
Statistics Canada survey, 53
Steffen, P., 99
Step counters and pedometers. *See* Pedometers and other step counters
Stevens, T.M., 76
Stoner, P., 99
Storm-van Essen, L., 57, 74
Strain gauge accelerometer, 84-85
Strazullo, P., 57
Suel, J., 57
Surgical patients, 20
Suter, E., 58
Sweden, questionnaire/interview-based studies in, 54, 57, 58
Swimming, 21
Symbols and abbreviations, 119-120

T

Tamura, T., 84-85
Task recording and analysis on computer (TRAC), 30-31
Taylor, H.L., 43, 44, 109
TBW (Total Body Water), 18
Tecumseh questionnaire/interview, 44-45, 62, 134-139, 145
compared to other methods, 46, 47, 87
compared to other questionnaires, 46, 52, 54
Telepedometer, 75
Ten Hoor, F., 88-89, 109
Thermic effect of food, 4
Thermodynamics, first law of, 6
Thivierge, M., 100
Thompson, P.D., 44
Thorland, W.G., 26
Time-lapse cinematography, 27
Time/motion studies. *See* Behavioral observation method
Togawa, T., 84-85
Torún, B., 26
Total Body Water (TBW), 18
Tour de France bicycle race, 20
TRAC (task recording and analysis on computer), 30-31
Triaxial accelerometers, 85-86, 88-89, 109, 110
Tri Trac-R3D, 89

U

Uniq Heart Watch, 100, 101-102
United States Employment Service, 44
U.S. National Health Survey, 53, 164-168
U.S. Public Health Service, 57
Units of measurement, 4, 120
University of Wisconsin, 30

V
Vaartiainen, E., 57
Validation, basic concerns summarized, 4-5
Van Cott, C.C., 101
Van Santen, E., 18
Van't Hof, M.A., 104
Van Waesberghe, F., 99
Ventilation method, 97-98
Verschuur, R., 57, 74
Videotaping, 11, 27, 107-108
Viikari, J., 57
Visscher, M.B., 17
Vitalog systems, 109
VO_2. *See* Oxygen consumption
VO_2max. *See* Aerobic power
Vollmer, J., 32
Volume conversion table, 120

W
Wade, M.G., 26
Wakabayashi, R., 84-85
Waldrop, M.F., 80, 86
Wallace, J.P., 56
Walter Reed Army Institute of Research, 86
Wanne, O., 57
Washburn, Richard A., 45, 60, 77, 79, 104
Webb, P., 7
Webster, J.G., 11
Weight conversion table, 120
Weir, J.B. de V., 10
Welford, N.T., 27
Welle, S., 19
Weller, G.M., 80
Westerterp, K.R., 19, 88-89, 105, 109
Williams, K.R., 11, 12
Wilson, M.F., 75
Wolff, H.S., 9-10, 27, 99
Wolffe, Joseph B., 3-4
Wong, W., 19
Woolfrey, J., 32
Work capacity. *See* Aerobic power
World Health Organization (WHO), 4
Wyborne, V.G., 83

Y
Yale Physical Activity Survey (YPAS), 55, 56, 62, 88, 175-180
Yamasa pedometer, 73, 74
Yasin, S., 50. *See also* British Civil Servant questionnaire
Yee, M.C., 109
YPAS (Yale Physical Activity Survey), 55, 56, 62, 88, 175-180

Z
Zehr, P., 102
Zuntz, Nathan, 9
Zutphen questionnaire, 55, 62, 172-174

About the Authors

Henry J. Montoye has specialized in the measurement of physical activity and energy expenditure since 1979. Henry is professor emeritus and former director of the Biodynamics Laboratory at the University of Wisconsin at Madison. He is a member and past president of the American College of Sports Medicine as well as a member of the American Physiological Society and the American Alliance for Health, Physical Education, Recreation and Dance. Henry is the author of more than 200 published research and scholarly articles. In his free time he enjoys skiing, sailing, bicycling, and golf.

Han C.G. Kemper is a specialist in the epidemiology of physical activity. He is head of the Department of Health Science at the Frije Universiteit in Amsterdam, Holland, where he serves as professor on the faculty of human movement sciences. Han is a fellow of the American College of Sports Medicine and a member of the European and North American Societies of Pediatric Exercise. In his leisure time he enjoys jogging, tennis, and following European soccer.

Since 1974 Wim H.M. Saris has researched human energy expenditure and daily physical activity. He is a professor in human nutrition and scientific director of the Nutrition Research Institute at the University of Limburg in Maastricht, Holland, where he also helped develop the Department of Human Biology and the Nutrition Research Institute. Wim is a member of the European Society of Pediatric Exercise and the European Association of the Study of Obesity. His pastimes include cycling, gardening, and oil painting.

Richard A. Washburn is a specialist in exercise epidemiology. He is an assistant professor in the Department of Kinesiology at the University of Illinois at Urbana-Champaign. Richard is a fellow in the American College of Sports Medicine and an author, coauthor, or presenter of more than 75 scholarly articles and abstracts on exercise epidemiology and related topics. In his free time he enjoys speedskating (ice and in-line), bicycling, tennis, and running.